Health and Safety Po

Health and Safety Pocket Book

Jeremy Stranks

AMSTERDAM • BOSTON • HEIDELBERG • LONDON • NEW YORK • OXFORD
PARIS • SAN DIEGO • SAN FRANCISCO • SINGAPORE • SYDNEY • TOKYO

Butterworth-Heinemann is an imprint of Elsevier

Elsevier Butterworth-Heinemann
Linacre House, Jordan Hill, Oxford OX2 8DP
30 Corporate Drive, Suite 400, Burlington, MA 01803

First edition 2006

British Library Cataloguing in Publication Data
A catalogue record for this book is available from the British Library

Library of Congress Cataloging-in-Publication Data
A catalog record for this book is available from the Library of Congress

ISBN-13: 978-0-7506-6781-4
ISBN-10: 0-7506-6781-8

For information on all Butterworth-Heinemann publications visit our
web site at http://books.elsevier.com

Typeset by Charon Tec Ltd, Chennai, India
www.charontec.com
Printed and bound in the UK

06 07 08 09 10   10 9 8 7 6 5 4 3 2 1

Working together to grow
libraries in developing countries

www.elsevier.com | www.bookaid.org | www.sabre.org

ELSEVIER   BOOK AID
International   Sabre Foundation

Contents

1(b)   The principal statutes   36

3(b)   Forms

Preface

Health and safety is a diverse subject covering many disciplines – law, engineering, human behaviour, safety management and occupational health – each of which is an area of study in its own right.

The *Health and Safety Pocket Book* has been written as a unique aid to health and safety practitioners and consultants, engineers, HR managers, lawyers and employee representatives. It should also be of use to those managers, across the full scope of industry and commerce, who may have specific responsibility for health and safety, together with those studying for specific qualifications in the discipline. The main objective is to provide a ready reference text on a wide range of issues, including the principal features of health and safety law, established management systems and sources of information.

A number of checklists, which are useful in the risk assessment process, have been incorporated, together with tables, figures and forms used on a regular basis. Specific parts include a glossary of commonly used terms, a summary of the legal requirements for documentation and record keeping, along with information on accredited health and safety training courses, professional organisations in health and safety and a breakdown of the legal requirements for the various industrial groups.

The A–Z arrangement within chapters and extensive cross-referencing makes the book easy to navigate. Individual references point the reader to the original legislation and more specialised reading.

I hope all who use this book will find it helpful.

Jeremy Stranks
November 2005

Abbreviations

ACOP	Approved Code of Practice
BS	British Standard
BSC	British Safety Council
CBI	Confederation of British Industry
CDM	Construction (Design and Management) Regulations
CE	Communité European
CHIP	Chemicals (Hazard Information and Packaging for Supply) Regulations
CORGI	Council for the Registration of Gas Installers
COSHH	Control of Substances Hazardous to Health
dB	Decibel
DSE	Display screen equipment
EA	Enforcing authority
EC	European Community
EHSRs	Essential health and safety requirements
ETBA	Energy Trace and Barrier Analysis
FPA	Fire Protection Association
HAZOPS	Hazard and Operability Studies
HFL	Highly flammable liquid
HMSO	Her Majesty's Stationery Office
HSC	Health and Safety Commission
HSE	Health and Safety Executive
HSWA	Health and Safety at Work etc. Act
HPZ	Hearing protection zone
IOELVs	Indicative Occupational Exposure Limit Values
ISO	International Standards Organisation
LEAV	Lower exposure action value
LEV	Local exhaust ventilation
LOLER	Lifting Operations and Lifting Equipment Regulations
LPG	Liquefied petroleum gas
Lx	Lux

MHSWR	Management of Health and Safety at Work Regulations
MORT	Management of Risk Tree
P	So far as is practicable
PHP	Personal hearing protection
PPE	Personal protective equipment
QSA	Quality Systems Auditing
RP	So far as is reasonably practicable
RoSPA	Royal Society for the Prevention of Accidents
RPE	Respiratory protective equipment
SCOEL	Scientific Committee on Occupational Exposure Limits
SHH	Substance(s) hazardous to health
THERP	Technique for Human Error Rate Probability
UEAV	Upper Exposure Action Value
WEL	Workplace Exposure Limit

PART 1
Health and Safety Law

1(a)
Legal background

Health and safety law covers many aspects involving people at work, including the civil and criminal liabilities of employers towards their employees and other persons.

The following topics are of particular significance in any consideration of the principal features of health and safety law.

Absolute (strict) liability

Certain duties under health and safety laws are of an absolute or strict nature. These duties are qualified by the terms 'shall' or 'must', such as the absolute duty on employers under the Management of Health and Safety at Work Regulations to undertake a suitable and sufficient risk assessment. Generally no defence is available although, when charged with an absolute offence, it may be possible to submit a plea in mitigation.

All reasonable precautions and all due diligence ('due diligence' defence)

Under certain Regulations, such as the Electricity at Work Regulations and the Control of Substances Hazardous to Health (COSHH) Regulations, an employer charged with an offence may be able to submit the defence that 'he took all reasonable precautions and exercised all due diligence to avoid the commission of the offence'.

To rely on this defence, the employer must establish that, on the balance of probabilities, he has taken *all* precautions that were reasonable and exercised *all* due diligence to ensure that these precautions were implemented in order to avoid such a contravention. It is unlikely that an employer could rely on this defence if:

(a) precautions were available which had not been taken; or
(b) that he had not provided sufficient information, instruction and training, together with adequate supervision, to ensure that the precautions were effective.

 1(a) Legal background
   Defences
 1(c) Principal regulations
   Control of Substances Hazardous to Health Regulations 2002
   Chemicals (Hazard Information and Packaging for Supply) Regulations 2002
   Pressure Systems Safety Regulations 2000

Approved Codes of Practice

The HSC is empowered to issue and approve Codes of Practice which accompany Regulations, e.g. *Workplace health, safety and welfare*, the Approved Code of Practice (ACOP) accompanying the Workplace (Health, Safety and Welfare) Regulations 1992.

An ACOP has limited legal status. Failure to comply with the recommendations in an ACOP may be used as evidence of failure to comply with a duty under Regulations, unless it can be shown that 'works of an equivalent nature' (which met the requirements but in a different way) had been undertaken.

 1(d) Approved Codes of Practice

Breach of statutory duty

In certain circumstances a breach of a statutory duty, which results in injury to a person of a class which the statute was designed to protect, will give the injured person a civil cause of action. The requirements which have to be satisfied before such a cause of action arises are:

(a) that the statutory provision, properly construed, was intended to protect an ascertainable class of persons of whom the claimant was one;

(b) that the provision has been broken;

(c) that the claimant had suffered damage of a kind against which the provision was designed to give protection; and

(d) that the damage was caused by the breach.

The claimant must prove his case by the ordinary standard of proof in civil actions. He must show at least that, on a balance of probabilities, the breach of duty caused, or materially contributed to, his injury.

Breaches of many Regulations, in addition to giving rise to criminal liability, also give rise to civil liability within the tort of breach of statutory duty.

 1(a) Legal background
Duty of care
Negligence

Burden of proof

This term applies to both criminal cases and civil claims.

Throughout criminal law the burden of proof of guilt that the accused person committed an offence rests with the prosecution, who must prove guilt 'beyond a reasonable doubt'. Section 40 of the HSWA makes the task of the prosecution easier by transferring the burden of proof to the accused. It is incumbent on the accused to show either that it was not 'practicable' or 'reasonably practicable' in the particular case to satisfy the particular duty or requirement. If the accused cannot discharge this duty, the case will be considered proved against him.

In civil claims, however, the claimant must show this proof of guilt on the part of the defendant 'on the balance of probabilities'.

Case law

Case law, fundamentally, is featured in the decisions of the criminal and civil courts, and based on the doctrine of judicial precedent. These doctrines are to be found in the various Law Reports, such as the All England Reports (AER) and the Industrial Relations Law Reports (IRLR). Case law is a self-endorsing process, perpetuated either by previous binding cases or by the interpretation of legislation.

The following features of a judgment are important:
- (a) the *ratio decidendi* (reason for deciding) – a statement of law based on an examination of the facts and the legal issues surrounding them; this is the most important part of a judgment and contains the actual binding precedent; and
- (b) the *obiter dicta* (words said by the way) – may contain a statement about the law which is not based on the facts of the case under review and which will not, therefore, be part of the decision; this is often held to be of persuasive authority.

 1(a) Legal background
Civil and criminal liability
Judicial precedent

Child

A person under compulsory school leaving age and under 16 years.

 1(c) Principal regulations
Management of Health and Safety at Work Regulations 1999

Civil and criminal liability

Breaches of health and safety law by employers and others can incur both criminal and civil liability.

Civil liability

Civil liability refers to the 'penalty' that can be imposed by a civil court, e.g. County Court, High Court, Court of Appeal (Civil Division) and the House of Lords.

A civil action generally involves a claim of negligence or breach of statutory duty by a claimant against a defendant. In such actions the claimant sues the defendant for a remedy that is beneficial to the claimant. In most cases, this remedy takes the form of damages, a form of financial compensation. In many cases, the claimant will agree to settle out of court.

Civil cases are decided on 'the balance of probabilities'.

Criminal liability

A crime is an offence against the State. Criminal liability refers to the duties and responsibilities of:
   (a) employers;
   (b) occupiers and controllers of premises;
   (c) manufacturers, designers and suppliers of articles and substances for use at work; and
   (d) employees,
under, principally, the HSWA, and regulations made under the HSWA, and to the penalties that can be imposed by the criminal courts, namely fines and imprisonment. The criminal courts involved are the Magistrates Courts, which handle the bulk of cases, the Crown Court, the Court of Appeal (Criminal Division) and the House of Lords.

Criminal law is based on a system of enforcement by the HSE, local authorities and fire authorities. A person charged with an offence is innocent until proved guilty 'beyond reasonable doubt'.

☞ **1(a) Legal background**
   Courts and tribunals
   Duty of care
   Negligence
☞ **1(b) The principal statutes**
   Health and Safety at Work etc. Act 1974
☞ **3(a) Tables and figures**
   Legal routes following an accident at work

Common law and statute law

Common law is the unwritten law in that it is not written down in Statutes and Regulations. It is, fundamentally, the body of accumulated case law (see separate entry) which is based on the decisions of the courts over many years, whereby precedents (see 'Judicial precedent') are established. It is of universal application and record in the various Law Reports. It is applicable to the decisions made by courts at their own level and in directions from superior courts.

Statute law, on the other hand, is the written law produced as a result of the parliamentary process. Statutes supersede all other forms of law and only Parliament can make, modify, revoke or amend statutes.

A statute may give the Minister or Secretary of State power to produce subordinate or delegated legislation, which generally takes the form of Regulations, e.g. the Control of Substances Hazardous to Health Regulations made under the HSWA.

☞ **1(a) Legal background**
   Civil and criminal liability
☞ **3(a) Tables and figures**
   Legal routes following an accident at work

Contractor

A person engaged to perform a certain task without direction from the person employing him, and implies a certain degree of independence from that person. The basic test of whether a person is an independent contractor is one of control over the undertaking of the work specified in the contract.

☞ **1(c) Principal regulations**
   Construction (Design and Management) Regulations 1994
   Construction (Health, Safety and Welfare) Regulations 1996
☞ **1(d) Approved Codes of Practice**
   Managing construction for health and safety
☞ **1(e) HSE guidance notes**
   Fire safety in construction: guidance for clients, designers and those managing and carrying out construction work involving significant risks
   Health and safety in construction
   Health and safety in excavations
   Health and safety in roof work
   Managing contractors
   The safe use of vehicles on construction sites
☞ *Appendix B: Documentation and record keeping requirements*

Contributory negligence

Where a person suffers damage, as the result partly through his own fault and partly through the fault of another person or persons, a claim in respect of that damage shall not be defeated by reason of the fault of the person suffering the damage, but the damages recoverable will be reduced to such extent as a court considers just and equitable having regard to the claimant's share in the responsibility for the damage.

☞ **1(a) Legal background**
   Duty of care
   Negligence

Controlling mind (*mens rea*)

 1(a) Legal background
Corporate liability

Corporate liability

Corporate liability refers to the liability of all those directing an undertaking, that is, the corporate body.

A corporate body, for example, the Board of Directors, chief executive, managing director, etc. of an organisation, may be liable for most criminal offences, providing a fine is specified for the offence, the offence is committed by a 'controlling mind', such as a managing director or chief executive, and is committed in the course or his corporate duties.

Under the HSWA, directors, managers, company secretaries and similar officers of the body corporate have both general and specific duties. Breaches of these duties can result in individuals being prosecuted.

Offences by bodies corporate (Sec 37(1))

Where an offence under any of the relevant statutory provisions committed by a body corporate is proved to have been committed with the consent or connivance of, or to have been attributable to any neglect on the part of, any director, manager, secretary or other similar officer of the body corporate or a person who was purporting to act in any such capacity, he as well as the body corporate shall be guilty of that offence and shall be liable to be proceeded against and punished accordingly.

 1(a) Legal background
Controlling mind (mens rea)
 1(b) The principal statutes
Health and Safety at Work etc. Act 1974

Courts and tribunals

There are two distinct systems whereby the courts deal with criminal offences and civil actions respectively. Some courts have both criminal and civil jurisdiction, however.

Criminal Courts

The Magistrates Courts (or Courts of Summary Jurisdiction) in England and Wales, and the Sheriff Court in Scotland, are the courts of first instance for all criminal offences. Lay Justices of the Peace (JPs) determine and sentence for the majority of the less serious offences. They also hold preliminary examinations into other offences to ascertain whether the prosecution can show a prima facie case on which the accused may be committed for trial at a higher court. The Sheriff Court performs a parallel function in Scotland, although procedures differ from those of the Magistrates Courts.

Serious indictable criminal charges and cases where the accused has the right to trial before a jury are heard on indictment in the Crown Court before a judge and jury. This court is empowered to impose unlimited fines and/or a maximum of two years imprisonment for health and safety-related offences. The Crown Court also hears appeals from the Magistrates Courts.

Civil Courts

County Courts operate on an area basis and deal in the first instance with a wide range of civil matters, such as claims for negligence. They are limited, however, in the remedies that can be applied. Cases are generally heard before a circuit judge or registrar, the latter having limited jurisdiction. A County Court judge can award compensation up to £50 000.

More important civil matters, because of the sums involved or legal complexity, will start in the High Court of Justice before a High Court judge. The High Court has three divisions:

(a) Queen's Bench – deals with contracts and torts; claims in excess of that within the County Court's power.

The Queen's Bench Division hears appeals on matters of law:

  (i) from Magistrates Courts and from the Crown Court on a procedure called 'case stated'; and

  (ii) from some tribunals, for example the finding of an employment tribunal on an enforcement notice under the HSWA;

It also has some supervisory functions over the lower courts and tribunals if they exceed their powers or fail to undertake their functions properly.

(b) Chancery – deals with matters relating to, for example, land, wills, bankruptcy, partnerships and companies;

(c) Family – deals with matters involving, for example, adoption of children, marital property and disputes.

The High Court, the Crown Court and the Court of Appeal are known as the Supreme Court of Judicature.

The Court of Appeal

The Court of Appeal has two divisions:

(a) the Civil Division, which hears appeals from the County courts and the High Court; and

(b) the Criminal Division, which hears appeals from the Crown Court.

The House of Lords

The Law Lords deal with matters of law only, following appeal from the Court of Appeal and, in restricted circumstances, from the High Court.

The European Court of Justice

This is the supreme law court whose decisions on the interpretation of European Community law are sacrosanct. These decisions are enforceable through the network of courts and tribunals in all Member States. The ECJ has jurisdiction in the following areas:

(a) Preliminary Ruling Jurisdiction – this enables the court to hear cases referred by the national courts of Member States in matters relating to the interpretation and application of Community law.

(b) Plenary Jurisdiction – this gives the court the right to award damages for unlawful acts committed by Community institutions.

(c) Contentious Jurisdiction – this gives the court the right to hear actions between Member States and Community institutions.

Cases can only be brought before the ECJ by organisations or individuals representing organisations.

The ECJ may also give advisory opinion to the Council of Ministers and the European Commission on legal matters.

Employment tribunals

Employment tribunals deal with many employment matters, including industrial relations issues and cases involving unfair dismissal, equal pay and sex discrimination.

Each tribunal consists of a legally qualified chairman appointed by the Lord Chancellor and two lay members, one representing employers and one from a trade union, selected from panels maintained by the Department of Employment following nominations from employers' organisations and trade unions.

When all three members of a tribunal are sitting, the majority view prevails.

Employment tribunals deal with the following health and safety-related issues:

(a) appeals against Improvement and Prohibition Notices served by the enforcement agencies;

(b) time off for the training of safety representatives;

   (c) failure of an employer to pay safety representatives for time off for undertaking their functions and for training;

   (d) failure of an employer to make a medical suspension payment; and

   (e) dismissal, actual or constructive, following a breach of health and safety law, regulation and/or term of employment contract.

Employment appeals tribunals, presided over by a judge, hear appeals on points of law from employment tribunals.

 **1(c) Principal regulations**

   Safety Representatives and Safety Committees Regulations 1977

 **3(a) Tables and figures**

   Legal routes following an accident at work

Damages

Civil liability may result in an award of damages for injury, disease or death at work in circumstances disclosing a breach of common law and/or statutory duty on the part of an employer/occupier of premises and arising out of and in the course of employment.

General damages relate to losses incurred after the hearing of an action, namely actual and probable loss of future earnings following an accident.

Special damages relate to quantifiable losses incurred before the hearing of the case, and consist mainly of medical expenses and loss of earnings.

In the case of fatal injury, compensation for death negligently caused is payable under the Fatal Accidents Act 1976, and a fixed lump sum is payable under the Administration of Justice Act 1982 in respect of bereavement.

1(a) Legal background
   Contributory negligence
   Duty of care
   Negligence

Defences

Civil actions

Where presented with a civil claim based on negligence or breach of statutory duty, a defendant may deny liability on the following grounds:

- (a) that the duty alleged to have been breached was never owed to the claimant in the first place;
- (b) that the nature of the duty was different from that pleaded by the defendant – that the duty was complied with;
- (c) that the breach of duty did not lead to the injury, damage or loss in question;
- (d) that the claimant was partly to blame i.e. was guilty of contributory negligence, which resulted in injury damage or loss.

 1(a) Legal background
   Duty of care
   Contributory negligence
   Negligence
   Res ipsa loquitur
   Volente non fit injuria

Criminal charges

Where charged with an offence under the HSWA or any of the relevant statutory provisions, a defendant may submit a defence on the basis that he had taken all 'practicable' or 'reasonably practicable' measures to comply with the requirement.

In other cases, Regulations such as the Pressure Systems Safety Regulations 2000, provide the defence of having taken all

reasonable precautions and exercised all due diligence to prevent the commission of an offence.

☞ **1(a) Legal background**
   All reasonable precautions and all due diligence
   Burden of proof
   Duties (hierarchy of)
☞ **1(b) The principal statutes**
   Health and Safety at Work etc. Act 1974
☞ **1(c) Principal regulations**
   Control of Substances Hazardous to Health Regulations 2002
   Chemicals (Hazard Information and Packaging for Supply)
      Regulations 2002
   Pressure Systems Safety Regulations 2000

Delegated legislation

A statute may delegate to the Minister or Secretary of State
power to make specific and detailed legislation on requirements
covered in the statute. Delegated or subordinate legislation is
exercised through Statutory Instruments (Sis) and generally takes
the form of Regulations, such as the Provision and Use of Work
Equipment Regulations and the Control of Substances Hazardous
to Health (COSHH) Regulations. Many Regulations are now introduced to meet the requirements of various European Directives.

☞ **1(b) The principal statutes**
   Health and Safety at Work etc. Act 1974
☞ **1(c) Principal regulations**
   Regulations made under the Health and Safety at Work etc.
      Act 1974

Disclosure of information

Section 28 of the HSWA requires that no person shall disclose any
information obtained by him or her as a result of the exercise of

any power conferred by Sections 14 or 20 (including, in particular, any information with respect to any trade secret obtained by him or her in any premises entered by him or her by virtue of any such power) except:

(a) for the purpose of his functions;
(b) for the purpose of any legal proceedings, investigation or inquiry, for the purpose of a report of any such proceedings or inquiry or of a special report made by virtue of Section 14; or
(c) with the relevant consent.

Information must not normally be disclosed except with the consent of the person providing it. Disclosure may be made in certain cases:

(a) for the purpose of any legal proceedings, investigation or inquiry held at the request of the HSC;
(b) with the relevant consent;
(c) for providing employees or their representatives with health and safety-related information.

 1(b) The principal statutes
Health and Safety at Work etc. Act 1974

Due diligence

 1(a) Legal background
All reasonable precautions and all due diligence
Defences

Duties (hierarchy of)

Duties on employers and others under health and safety law may be absolute or strict, or qualified by the terms 'so far as is practicable' or 'so far as is reasonably practicable'.

Absolute requirements

Where risk of death, injury or disease is inevitable if health and safety requirements are not complied with, a statutory duty may well be strict or absolute. An example of an absolute duty is to be found in Regulation 5(1) of the Provision and Use of Work Equipment Regulations which states:

> Every employer shall ensure that work equipment is so constructed or adapted as to be suitable for the purpose for which it is to be used or provided.

Absolute duties are qualified by the term 'shall' or 'must' and there is little or no defence available when charged with such an offence.

'Practicable' requirements

A duty qualified by the term 'so far as is practicable' implies that if in the light of current knowledge or invention, or in the light of the current state of the art, it is feasible to comply with the requirement then, irrespective of cost or sacrifice involved, such a requirement must be complied with. [*Schwalb v Fass H & son (1946) 175 LT 345*]

'Practicable' means more than physically possible and implies a higher duty of care than a duty qualified by 'so far as is reasonably practicable'.

'Reasonably practicable' requirements

'Reasonably practicable' is a narrower term than 'physically possible' (i.e. 'practicable') and implies that a computation must be made in which the quantum of risk is placed in one scale and the sacrifice involved in the measures necessary for averting the risk is placed in the other. If it can be shown that there is a gross disproportion between these two factors, i.e. the risk being insignificant in relation to the sacrifice, then a defendant discharges the onus upon himself. [*Edwards v National Coal Board (1949) 1 AER 743*]

Most duties under the HSWA are qualified by the term 'so far as is reasonably practicable'.

 1(b) The principal statutes
Health and Safety at Work etc. Act 1974
 1(c) Principal regulations
Construction (Design and Management) Regulations 1994
Control of Substances Hazardous to Health Regulations 2002
Management of Health and Safety at Work Regulations 1999
Manual Handling Operations Regulations 1992
Provision and Use of Work Equipment Regulations 1998

Duty holders

A term used in certain Regulations, such as the Electricity at Work Regulations and the Construction (Design and Management) Regulations, specifying classes of person on whom duties are imposed, such as employers, self-employed persons, clients and contractors.

 1(c) Principal regulations
Construction (Design and Management) Regulations 1994
Electricity at Work Regulations 1989

Duty of care

The common duty of care between one person and another, such as an employer and employee, occupier of premises and visitor, or the manufacturer of a product and the user of that product, is a key principle of common law. Fundamentally, everyone owes a duty to everyone else to take reasonable care so as not to cause them foreseeable injury.

The effect of this requirement is that if an employer knows of a risk to the health and safety of his employees, or ought, in the light of knowledge current at that time, to have known of the existence of such a risk, he will be liable if an employee is injured or suffers death as a result of that risk, or if the employer failed to take reasonable care to avoid the risk arising.

 1(a) Legal background
Common law and statute law
Contributory negligence
Negligence

Employers' duties (common law)

The duties of an employer under the common law were established in general terms in *Wilson's & Clyde Coal Co. Ltd. v English (1938) AC 57 2 AER 68*. Under the common law all employers must provide and maintain:
  (a) a safe place of work with safe means of access to and egress from same;
  (b) safe appliances and equipment and plant for doing the work;
  (c) a safe system for doing the work; and
  (d) competent and safety-conscious personnel.

 1(a) Legal background
Case law
Common law and statute law
Contributory negligence
Duty of care
Negligence

Employers' liability

Employers are vicariously liable for the actions of their employees. This liability must be insured against under the Employers' Liability (Compulsory Insurance) Act 1969. Employers cannot contract out of this liability as such practices are prohibited by the Law Reform (Personal Injuries) Act 1948 and the Unfair Contract Terms Act 1977.

 1(a) Legal background
  Vicarious liability
 1(b) The principal statutes
  Health and Safety at Work etc. Act 1974
  Unfair Contract Terms Act 1977

Enforcement arrangements – powers of inspectors

Enforcement authorities, such as the HSE, appoint inspectors who have specific powers under the HSWA. These powers are specified in section 20 of the HSWA thus:

(a) to enter premises at any reasonable time and, where obstruction is anticipated, to enlist the support of a police officer;

(b) on entering premises:
  (i) to take with him any person duly authorised by his enforcing authority; and
  (ii) any equipment or materials required for any purpose for which the power of entry is being exercised;

(c) to make such examination and investigation as may be necessary;

(d) to direct that premises or any part of such premises, or anything therein, shall remain undisturbed for so long as is reasonably necessary for the purposes of examination or investigation;

(e) to take such measurements and photographs and make such recordings as he considers necessary for the purposes of any examination or investigation;

(f) to take samples of any articles or substances found in any premises, and of the atmosphere in or in the vicinity of such premises;

(g) where it appears to him that any article or substance has caused or is likely to cause danger to health or safety, to cause it to be dismantled or subjected to any process or test;

(h) to take possession of any article or substance and to detain same for so long as is necessary:

　(i) to examine same;

　(ii) to ensure it is not tampered with before his examination is completed; and

　(iii) to ensure it is available for use as evidence in any proceedings for an offence under the relevant statutory provisions;

(i) to require any person whom he has reasonable cause to believe to be able to give any information relevant to any examination or investigation to answer such questions as the inspector thinks fit and to sign a declaration of the truth of his answers;

(j) to require the production of, inspect and take copies of, any entry in:

　(i) any books or documents which by virtue of the relevant statutory provisions are required to be kept; and

　(ii) any other books or documents which it is necessary for him to see for the purposes of any examination or investigation;

(k) to require any person to afford him such facilities and assistance with respect to any matter or things within that person's control or in relation to which that person has responsibilities as are necessary to enable the inspector to exercise any of the powers conferred on him by this section; and

(l) any other power which is necessary for the purpose of carrying into effect the relevant statutory provisions.

Inspectors may serve two types of notice.

Improvement Notice

If an inspector is of the opinion that a person:

(a) is contravening one or more of the relevant statutory provisions; or

(b) has contravened one or more of those provisions in circumstances that make it likely that the contravention
   will continue or be repeated,

he may serve on him an Improvement Notice stating that he is
of the opinion, specifying the provision or provisions as to
which he is of that opinion, giving particulars of the reasons
why he is of that opinion, and requiring the person to remedy
the contravention or, as the case may be, the matters occa-
sioning it within such period (ending not earlier than the
period in which an appeal against the notice can be brought
under section 24) as may be specified in that notice.

Prohibition Notices

Where an inspector is of the opinion that a work activity involves
or will involve a risk of serious personal injury, he may serve a
Prohibition Notice on the person in control of that activity.

A Prohibition Notice shall:
   (a) state that the inspector is of the said opinion;
   (b) specify the matters which in his opinion give or, as the
      case may be, will give rise to the said risk;
   (c) where in his opinion any of those matters involves or, as
      the case may be, will involve a contravention of any of
      the relevant statutory provisions, state that he is of that
      opinion, specify the provision or provisions as to which
      he is of that opinion, and give particulars of the reasons
      why he is of that opinion;
   (d) direct that the activities to which the notice relates shall
      not be carried on by or under the control of the person
      on whom the notice is served unless the matters speci-
      fied in the notice have been remedied.

A direction given in a Prohibition Notice shall take immediate effect if the inspector is of the opinion, and states it, that
the risk of serious personal injury is or, as the case may be, will
be imminent, and shall have effect at the end of a period
specified in the notice in any other case (deferred Prohibition
Notice).

☞ **1(b) The principal statutes**
　　Health and Safety at Work etc. Act 1974
☞ **3(a) Tables and figures**
　　Legal routes following an accident at work
☞ **3(b) Forms**
　　Improvement Notice
　　Prohibition Notice

Enforcement authorities

The enforcing authorities for the HSWA and other health and safety legislation are:
　　(a) the Health and Safety Executive (HSE), which is split into a number of specific inspectorates, e.g. Nuclear Installations, Agriculture, and National Industrial Groups (NIGs);
　　(b) local authorities, principally through their environmental health departments; and
　　(c) fire and rescue authorities, for certain fire-related legislation affecting workplaces.

☞ **1(b) The principal statutes**
　　Health and Safety at Work etc. Act 1974

Guidance notes (HSE)

The HSE produce guidance on a very wide range of matters. Guidance notes are issued in six specific series:
　　(a) general safety;
　　(b) chemical safety;
　　(c) environmental hygiene;
　　(d) medical;
　　(e) plant and machinery;
　　(f) health and safety.
Guidance notes have no legal status and are not, generally, admissible as evidence in criminal proceedings.

Guidance notes are also issued with Regulations e.g. *Personal protective equipment* accompanying the Personal Protective Equipment at Work Regulations 1992.

 1(e) HSE guidance notes

Indictable offences

Where there is sufficient evidence, certain offences can, on the decision of a Magistrates Court, be subject to committal pro-ceedings through issue of an indictment, whereby an offender is committed to a Crown Court for trial.

Certain offences triable only on indictment are:
   (a) breaching any of the relevant statutory provisions; and
   (b) acquiring, or attempting to acquire, possessing or using an explosive article or substance.

 1(a) Legal background
   Burden of proof
   Courts and tribunals
   Delegated legislation
   Duties (hierarchy of)
   Enforcement arrangements
   Summary offences
 1(b) The principal statutes
   Health and Safety at Work etc. Act 1974

Judicial precedent

Judicial precedent is defined as 'a decision of a tribunal to which some authority is attached'. A precedent may be authoritative or persuasive.

Authoritative precedents

These are the decisions which judges are bound to follow. There is no choice in the matter. A lower court, for example, is bound by the previous decision of a higher court.

Persuasive precedents

These are decisions which are not binding on a court but to which a judge will attach some importance. For example, decisions given by the superior courts in Commonwealth countries will be treated with respect in the English High Court.

 1(a) Legal background
 Burden of proof
 Case law
 Defences

Negligence

'Negligence' is defined as 'careless conduct injuring another'.

In a civil action for negligence, three specific facts must be proved by the claimant:

 (a) a duty of care is owed;
 (b) there has been a breach of that duty; and
 (c) injury, damage and/or loss has been sustained by the claimant as a result of that breach.
[*Lochgelly Iron & Coal Co Ltd v M'Mullen (1934) AC 1*]

 1(a) Legal background
 Breach of statutory duty
 Defences
 Duty of care
 Contributory negligence
 Courts and tribunals
 Employers' duties (common law)

Neighbour Principle
Res ipsa loquitur
Torts
Vicarious liability
Volente non fit injuria

Neighbour Principle

This principle was established in *Commissioner for Railways v McDermott (1967) 1 AC 169*. Lord Gardner explained the position with regard to occupation of premises thus:

> Occupation of premises is a ground of liability and not a ground of exemption of liability. It is a ground of liability because it gives some control over and knowledge of the state of the premises, and it is natural and right that the occupier should have some degree of responsibility for the safety of persons entering his premises with his permission. There is proximity between the occupier and such persons and they are his neighbours. Thus arises a duty of care.

 1(a) Legal background
Duty of care
Employers' duties (common law)
Employers' liability
Negligence
Occupiers' liability

No-fault liability

This term implies there is no requirement to establish fault or intent in that a claimant is automatically compensated for injury, damage or loss by the state or some other organisation. It thus dispenses with or disregards the common law and statutory elements of liability.

Occupiers' liability

People who occupy land and premises, such as private individuals, local authorities, organisations, companies and shop keepers, owe a common duty of care to all persons who may visit their land or premises. Moreover, anyone who is injured whilst visiting or working on land or premises may be in a position to sue the occupier for damages, even though the injured person may not be his employee.

[Lord Gardner in *Commissioner of Railways v McDermott (1967) 1 AC 169*]

The law covering this area of civil law is the Occupiers' Liability Acts 1957 and 1984.

Occupiers' Liability Act 1957

Under the Occupiers' Liability Act (OLA) an occupier owes a common duty of care to all lawful visitors. This common duty of care is defined as 'a duty to take such care as in all the circumstances is reasonable to see that the visitor will be reasonably safe in using the premises for the purposes for which he is invited or permitted by the occupier to be there'.

Under the OLA, occupiers have a duty to erect notices warning visitors of imminent danger. The display of a warning notice does not, however, absolve an occupier from liability unless, in all the circumstances, it was sufficient to enable the visitor to be reasonably safe.

However, whilst an occupier may have displayed a notice under the provisions of the OLA, the chance of avoiding liability is not permitted as a result of the Unfair Contract Terms Act 1977. This Act states that it is not permissible to exclude liability for death or injury due to negligence by a contract or by a notice displayed in accordance with the OLA.

The OLA 1984 went on to clarify the situation relating to the display of warning notices. Whilst a duty may be discharged by the

Enforcement arrangements
Indictable offences
Summary offences
 1(b) The principal statutes
Health and Safety at Work etc. Act 1974

Relevant statutory provisions

The HSWA is an enabling Act comprising, in the main, general duties on specific groups, e.g. employers, manufacturers of articles for use at work. Section 15 of the HSWA enables the Secretary of State for Employment to make Regulations. The HSWA and Regulations are deemed to be 'the relevant statutory provisions'.

Schedule 1 of the HSWA defines 'the relevant statutory provisions' as:
(a) Part 1 of the HSWA;
(b) regulations made under Part 1;
(c) the Acts contained in Schedule 1 of the HSWA e.g. Factories Act 1961; and
(d) any regulations made under the above Acts.

 1(a) The principal statutes
Health and Safety at Work etc. Act 1974
 1(c) Principal regulations
Management of Health and Safety at Work Regulations 1999

Res ipsa loquitur

A term meaning 'the facts speak for themselves' or 'the matter itself speaks', a phrase used in actions for injury occasioned by negligence where no proof is required of negligence beyond the accident itself.

 1(a) Legal background
Negligence

Stare decisis

A term meaning 'let the decision stand', implying the need to stick to the decisions of past cases. *Stare decisis* is the basis for the doctrine of binding precedent.

Statement of claim

A statement, outlining the allegations against a defendant, and previously delivered to the defendant prior to civil proceedings.

 1(a) Legal background
   Civil and criminal liability
   Courts and tribunals

Statement of health and safety policy

A formal document required to be prepared, and regularly revised, by an employer under the HSWA. The principal features of such a statement are:
   (a) statement of intent, which outlines the organisation's overall philosophy in relation to the management of health and safety, including objectives for ensuring legal compliance;
   (b) details of the organisation, which should include the chain of command, accountability and responsibility for health and safety; and
   (c) the arrangements, which include the procedures and systems for monitoring performance.

 1(b) The principal statutes
   Health and Safety at Work etc. Act 1974
 1(e) HSE guidance
   Writing your health and safety policy statement: a guide to preparing a safety policy for a small business

Statutes and Regulations

Statutes (Acts of Parliament) are the principal written laws arising from the parliamentary process. Only Parliament can make,
modify, amend or revoke statutes.

A statute frequently gives the Minister of Secretary of State power
to make Regulations ('delegated' or 'subordinate' legislation).
The majority of Regulations, such as the Provision and Use of
Work Equipment Regulations, are based on the requirements
of European Directives, and are listed numerically as Statutory
Instruments.

 1(b) The principal statutes
 1(c) Principal regulations

Strict liability (see Absolute liability)

Summary offences

These are offences which may be dealt with in a court of summary jurisdiction, e.g. a Magistrates Court. Such offences are
mainly of a minor nature.

 1(a) Legal background
        Courts and tribunals
 1(b) The principal statutes
        Health and Safety at Work etc. Act 1974

Temporary employment

Temporary employment arises where an employee of one
organisation may be hired out or seconded to another employer
to undertake a specific task. The test as to whether an employee
has been 'temporarily employed' is based on the 'control test',

namely the extent of the control the temporary employer can exert in that employee's actual undertaking of the specific task involved.

[*Mersey Docks & Harbour Board v Coggins and Griffiths (Liverpool) Ltd (1974) AC1*]

Regulation 12 of the Management of Health and Safety at Work Regulations, *Persons working in host employers' or self-employed persons' undertakings*, covers the criminal law requirements of temporary employment.

 1(c) Principal regulations
   Management of Health and Safety at Work Regulations 1999

Torts

A tort is defined as a 'civil wrong'. The three principal torts are negligence, nuisance and trespass.

The rule of common law is that everyone owes a duty to everyone else to take reasonable care so as not to cause them foreseeable injury. Tortious liability arises from the breach of a duty primarily fixed by the law. Such duty is towards persons generally and its breach is redressible by an action for unliquidated damages.

 1(a) Legal background
   Breach of statutory duty
   Defences
   Duty of care
   Negligence

Vicarious liability

The doctrine of vicarious liability is based on the fact that if an employee, whilst acting in the course of his employment, negligently injures another employee, or the employee of a contractor working on the premises, or even a member of the public, the employer, rather than the employee, will be liable for that injury. Vicarious liability rests on the employer simply as a result of the fact that he is the employer and is deemed to have ultimate control over the employee in a 'master and servant' relationship.

The key to liability is that the accident causing the injury, disease or death arises, firstly, *out of* and, secondly, *in the course of*, employment. This does not normally include the time travelling to and from work, although it would apply if the mode of transport was within the employer's control or was provided by the employer or by arrangement with the employer.

 1(a) Legal background
Duty of care
Negligence

Volente non fit injuria

This term means 'to one who is willing, no harm is done', more commonly referred to as 'voluntary assumption of risk'. As such it is a complete defence and no damages would be payable to a claimant in respect of a claim for negligence.

However it is a very limited defence based on the employee voluntarily assuming a risk as part of his normal work activities.

[*Smith v Baker & Sons (1891) AC 305*]

 1(a) Legal background
Defences
Negligence

1(b)
The principal statutes

Administration of Justice Act 1982

Liabilities of persons
- Where death is caused by any wrongful act, neglect or default which is such as would (if death had not ensued) have entitled the person injured to maintain an action and recover damages, the person who would have been liable if death had not ensued, shall be liable for damages.
- A lump sum is payable to defendants
- Subsequent remarriage, or the prospect of remarriage, of a dependant must not be taken into account in assessing fatal damages.

Disability Discrimination Act 1995

Responsible for enforcement
   Tribunals
Level of duty
   Absolute
Duties of employers
- Make 'reasonable adjustments' where working arrangements and/or the physical features of a workplace cause a substantial disadvantage for a disabled person, including:
   - altering working hours
   - allowing time off for rehabilitation or treatment
   - allocating some of the disabled person's duties to someone else
   - transferring the disabled person to another vacancy or another place of work
   - giving or arranging training
   - providing a reader or interpreter

- ○ acquiring or modifying equipment or reference manuals
- ○ adjusting the premises.
- In considering physical features of a workplace, take into account:
  - ○ those arising from the design or construction of a building
  - ○ exits or accesses to buildings
  - ○ fixtures, fittings, furnishings, equipment or materials
  - ○ any other physical element or quality of land or the premises.
- Make reasonable adjustments to:
  - ○ arrangements to determine who is offered employment
  - ○ terms and conditions or other arrangements on which employment, promotions, transfer, training or other benefits are offered.

Employers' Liability (Compulsory Insurance) Act 1969

Responsible for enforcement
   HSE, local authorities
Level of duty
   Absolute
Duties of employers

- Insure against claims by employees suffering personal injury, damage or loss
- Display Certificate of Insurance conspicuously at the workplace
- Disclose all appropriate information to an insurer
- Ensure policy is an approved policy by virtue of the Employers' Liability (Compulsory Insurance) General Regulations 1971
- Make policy and certificate available to an enforcement officer.

Duties of insurers
- Ensure that the policy states that any person under a contract of service or apprenticeship who sustains injury, disease or death caused during the period of insurance and arising out of the course of employment will be covered for any legal liability on the part of his employer to pay compensation.

Employers' Liability (Defective Equipment) Act 1969

Liabilities of employers
- Strictly liable for injuries to employees caused by defective equipment, where the defect is wholly or in part the result of manufacture, that is, by a third party
- An injury suffered by an employee is to be attributable to negligence by an employer in the following situations where
    - an employee suffers personal injury (including death) in the course of employment in consequence of a defect in equipment
    - the equipment was provided by the employer for use in the employer's business
    - the defect is attributable, wholly or in part, to the fault of a third party, whether identified or not, such as a manufacturer, supplier, distributor or importer
- Strictly liable for defects in manufacture and supply.

Employment Rights Act 1996

Duties of employers with respect to 'health and safety cases'
- An employee has the right not to be subjected to any detriment by any act, or any deliberate failure to act, by his employer done on the grounds that:
    - (a) having been designated by the employer to carry out activities in connection with preventing or reducing risks to health and safety at work, the employee carried out (or proposed to carry out) any such activities;

    (b) being a representative of workers on matters relating to health and safety at work or a member of a safety committee:

       (i) in accordance with arrangements established under or by virtue of any enactment; or

      (ii) by reason of being acknowledged as such by the employer, the employee performed (or proposed to perform) any functions as such a representative or a member of such a committee;

    (c) being an employee at a place where:

       (i) there was no such representative or safety committee; or

      (ii) there was such a representative or safety committee but it was not reasonably practicable for the employee to raise the matter by those means, he brought to his employer's attention, by reasonable means, circumstances connected with his work which he reasonably believed were harmful or potential harmful to health or safety;

    (d) in circumstances of danger which the employee reasonably believed to be serious and imminent and which he could not reasonably have been expected to avert, he left (or proposed to leave) or (while the danger persisted) refused to return to his place of work or any dangerous part of his place of work; or

    (e) in circumstances of danger which the employee reasonably believed to be serious and imminent, he took (or proposed to take) appropriate steps to protect himself or other persons from danger.

- For the purposes of sub-section (e) above, whether steps which an employee took (or proposed to take) were appropriate is to be judged by reference to all the circumstances including, in particular, his knowledge and the facilities and advice available to him at the time.

- An employee is not to be regarded as having been subjected to any detriment on the ground specified in sub-section (e) above if the employer shows that it was (or would have been) so negligent for the employee to take the steps which he took (or proposed to take) that a reasonable employer might have treated him as the employer did.

Fatal Accidents Act 1976

Liabilities of persons
- In the case of fatal injury, compensation is payable to the dependants for death negligently caused in respect of financial loss suffered by them as a result of the death.

Health and Safety at Work etc. Act 1974

Responsible for enforcement
  HSE, local authorities
Level of duty
  Reasonably practicable
Duties of employers to employees
- General duty to ensure safety, health and welfare of all persons at work
- Extended duties in respect of, provision of:
  - safe plant and systems of work
  - arrangements for use, handling, storage and transport of articles and substances
  - information, instruction, training and supervision
  - maintenance of workplace plus safe access and egress
  - maintenance of a safe and healthy working environment with adequate welfare amenities
- Prepare and promote a Statement of Health and Safety Policy (A)
- Consult with appointed safety representatives.
Duties of employers to non-employees
- Conduct of undertaking as to protect non-employees
- Ensure provision of information to non-employees on hazards and precautions.
Duties of occupiers/controllers of premises
- Ensure provision of safe premises, access and egress, plant and substances in the premises or provided for use.

Duties of designers, manufacturers, etc. of articles for use at work
- Ensure safe design and construction of safe articles
- Ensure testing and examination of articles
- Ensure provision of information about articles
- Undertake research to minimise risks.

Duties of erectors, installers, etc.
- Ensure safe erection/installation.

Duties of manufacturers, etc. of substances for use at work
- Ensure safety and freedom from health risks of substances
- Carry out testing and examination of substances
- Make available adequate information about results of tests.

Duties of employees
- Take reasonable care of themselves and others (A)
- Co-operate with employer with respect to duties imposed on him (A).

Duties of all persons
- Not to intentionally or recklessly interfere with or misuse anything provided in the interests of health, safety or welfare in pursuance of the relevant statutory provisions.

Penalties
Maximum £20 000 for breach of section 2–6 (Magistrates Court)
Maximum £5000 for other offences (Magistrates Court)
Unlimited fines in higher courts
Imprisonment for breach of improvement notice and/or prohibition notice.

 3(c) Forms
Improvement notice
Prohibition notice

Occupiers' Liability Act 1957

Duties of occupiers
- Owe a common duty to take such care as in all the circumstances of the case is reasonable to see that a visitor will be

reasonably safe in using the premises for the purposes for which he is invited or permitted by the occupier to be there
- Applies to all persons lawfully on the premises in respect of dangers due to the state of the premises or to things done or omitted to be done on them
- Must erect notices warning visitors of any imminent danger
- Be prepared for children to be less careful than adults.

 1(a) Legal background
Duty of care
Occupiers' liability

Occupiers' Liability Act 1984

Duties of occupiers
- Owe a common duty of care in respect of trespassers, namely persons who may have lawful authority to be in the vicinity or not, who may be at risk of injury on an occupier's premises
- Display notices where there may be a risk to simple tres-passers and enforce the requirements of such notices.

 1(a) Legal background
Duty of care
Negligence
Occupiers' liability

Social Security Act 1975

Duties of employers
- Must provide and keep readily available an accident book in an approved form in which the appropriate particulars of all accidents can be recorded
- The appropriate particulars include:
  - name and address of the injured person
  - date and time of the accident

- ○ the place where the accident happened
- ○ the cause and nature of the injury
- ○ the name and address of any third party giving the notice
- Must retain such books, when completed, for three years after the date of the last entry
- Must investigate all accidents of which notice is given by employees (in the Accident Book) and must record any variations between the findings of this investigation and the particulars given in the notification
- Must, on request from the Department of Social Security (DSS), furnish the DSS with such information as may be required relating to accidents in respect of which benefit may be payable.

Duties of employees
- Must notify their employer of any accident resulting in personal injury in respect of which benefit may be payable; notification may be given by a third party if the employee is incapacitated
- Must enter the appropriate particulars of all accidents in an accident book; entry in the accident book may be made by a third party if the employee is incapacitated.

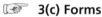 **3(c) Forms**

Accident book

Unfair Contract Terms Act 1977

Duties of persons
- Cannot, by reference to any contract term or to a notice given to persons, generally exclude or restrict liability for death or a personal injury resulting from negligence.

1(c)
Principal regulations

Building Regulations 2000

Responsible for enforcement
  Local authorities
Level of duty
  Absolute
Duties of persons
- Building work must be carried out with adequate and proper materials and in a workmanlike manner
- When intending to carry out building work or to make a material change of use shall:
  - give notice to the local authority
  - deposit full plans with the local authority
- A building notice shall state the name and address of the person intending to carry out the work and shall be signed by him and shall contain or be accompanied by:
  - a statement
  - a description of the proposed building work or material change of use
  - particulars of the location of the building
- In the case of an erection of a building, a building notice shall be accompanied by:
  - a plan showing the location of the building in relation to other buildings and streets
  - a statement specifying the number of storeys
  - particulars of provision to be made for drainage
- Specific requirements apply in the case of building work which involves the insertion of insulating material into the cavity walls of a building and work involving the provision of a hot water storage system

- Shall give the local authority notice:
    - prior to commencing building work and not cover up any foundation work or drainage work
    - after work has been completed
    - prior to occupation of a building
- Calculate the energy rating of a dwelling and give notice of that rating to the local authority

Duties of local authority

- Give a completion certificate on receipt of notification that building work has been completed or that a building has been partly occupied before completion
- Where satisfied that the relevant requirements of Schedule 1 specified in the certificate have been satisfied, give a certificate to that effect
- Make such tests of drains or private sewers as may be necessary
- Take samples of building materials in the carrying out of building work
- Take action in the case of unauthorised building work.

Schedule 1 – Requirements

Schedule 1 lays down requirements with regard to:

- A – Structure
    - Loading
    - Ground movement
    - Disproportionate collapse
- B – Fire safety
    - Means of warning and escape
    - Internal fire spread (linings)
    - Internal fire spread (structure)
    - External fire spread
    - Access and facilities for the fire service
- C – Site preparation and resistance to moisture
    - Preparation of site
    - Dangerous and offensive substances
    - Subsoil drainage
    - Resistance to weather and ground moisture
- D – Toxic substances
    - Cavity insulation

- E – Resistance to the passage of sound
  - Airborne sound (walls)
  - Airborne sound (floors and stairs)
  - Impact sound (floors and stairs)
- F – Ventilation
  - Means of ventilation
  - Condensation in roofs
- G – Hygiene
  - Sanitary conveniences and washing facilities
  - Bathrooms
  - Hot water storage
- H – Drainage and waste disposal
  - Foul water drainage
  - Cesspools, septic tanks and settlement tanks
  - Rainwater drainage
  - Solid waste storage
- J – Heat producing appliances
  - Air supply
  - Discharge of products of combustion
  - Protection of building
- K – Protection from falling, collision and impact
  - Stairs, ladders and ramps
  - Protection from falling
  - Vehicle barriers and loading bays
  - Protection from collision with open windows etc.
  - Protection against impact from and trapping by doors
- L – Conservation of fuel and power
- M – Access and facilities for disabled people
  - Definition of 'disabled people'
  - Access and use
  - Sanitary conveniences
  - Audience or spectator seating
- N – Glazing – Safety in relation to impact, opening and cleaning
  - Protection against impact
  - Manifestation of glazing
  - Safe opening and closing of windows etc.
  - Safe access for cleaning windows etc.

Chemicals (Hazard Information and Packaging for Supply) Regulations 2002

Responsible for enforcement
    HSE, local authorities
Level of duty
    Absolute
Defence
    All reasonable precautions and all due diligence
Duties of suppliers
- Not supply a substance or preparation dangerous for supply (SPDS) unless it has been classified
- Provide recipient of SPDS with a safety data sheet
- Keep safety data sheet up to date and revise as necessary
- Ensure substance is not advertised unless mention made in the advertisement of hazards presented by the substance
- Supply in a package suitable for that purpose
- Comply with labelling requirements for packages and for certain preparations, including methods of marking and labelling
- In the case of specified substances and preparations, the provision of child-resistant fastenings and tactile warning devices to receptacles
- Retain classification data for SPDS, making copies available to the appropriate enforcing authority
- Notify constituents of certain preparations dangerous for supply to the Poisons Advisory Centre.

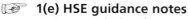 **1(e) HSE guidance notes**
    Approved Classification and Labelling Guide
    CHIP for everyone
 **3(a) Tables and figures**
    Categories of danger
    Safety data sheets

Children (Protection at Work) Regulations 1997

Responsible for enforcement
   HSE, local authorities
Level of duty
   Absolute
Duties of employers
- Must not employ a child below the minimum age (14 years) in any work other than as employee of his parent or guardian in light agricultural or horticultural work on an occasional basis
- Must not employ a child on anything other than 'light work', i.e. work which does not jeopardise a child's safety, health, development, attendance at school or participation in work experience
- May employ children over the age of 13 years in categories of light work specified in local authority byelaws
- Must not employ a child over the age of 14 years beyond the specified hours, including the specified rest periods
- Must provide at least one two-week period in a child's school holidays free from any employment
- Must not allow a child going abroad for the purposes of performing for profit, for the purposes of taking part in sport or working as a model in circumstances where payment is made, without a local authority licence.

Confined Spaces Regulations 1997

Responsible for enforcement
   HSE, local authorities
Level of duty
   Absolute
Duties of employers
- Avoid entry to confined spaces wherever possible, for example, doing the work from outside

- Follow a safe system of work if entry is unavoidable
- Put in place adequate emergency arrangements before work starts, which will also safeguard rescuers.

 1(d) Approved codes of practice
   Safe work in confined spaces
 1(e) HSE guidance notes
   Safe work in confined spaces

Construction (Design and Management) Regulations 1994

Responsible for enforcement
   HSE
Level of duty
   Absolute
Duties of clients

- Not to appoint a person as his agent unless reasonably satisfied as to his competence to perform the duties imposed on a client
- Make a declaration to the HSE of the appointed agent's name and address
- Appoint a competent planning supervisor and principal contractor
- Ensure construction phase of any project does not start unless a health and safety plan has been prepared (RP)
- Provide planning supervisor with information relevant to his functions about the state or condition of any premises at or on which construction work is or is intended to be carried out
- Take steps to ensure information in the health and safety file is kept available for inspection by any person who may need information
- When disposing of interest in a property, to transfer the health and safety file to the new owner.

Duties of developers

- In the case of domestic clients, to act on behalf of same to ensure compliance with regulations.

Duties of designers

- Ensure that any design he prepares includes among the design considerations the need:
  - to avoid foreseeable risks
  - to combat risks at source
  - to give priority to measures which will protect all persons at work
- Ensure that the design includes adequate information about any aspect of the project which may affect health and safety
- Co-operate with the planning supervisor and any other designer to enable compliance with the relevant statutory provisions
- Provide adequate information where risks cannot be avoided and alert the client to his duties.

Duties of planning supervisors

- Overall responsibility for co-ordinating the health and safety aspects of the design and planning stage of a project
- Ensure notice of project is given to the HSE
- Ensure design of any structure takes account of the three design considerations above (RP)
- Ensure co-operation between designers
- Provide advice to client to enable him to comply with his duties
- Ensure health and safety file is prepared in respect of each structure in the project, reviewing same when necessary
- Ensure health and safety file is delivered to the client on completion of construction work
- Ensure a health and safety plan in respect of a project has been prepared within the specified time.

Duties of principal contractors

- Ensure construction phase health and safety plan is pre-pared and contains prescribed information
- Take reasonable steps to ensure co-operation between contractors to enable compliance with the relevant statu-tory provisions

- Co-ordinate the activities of contractors to ensure compliance with the relevant statutory provisions
- Ensure any contractor complies with any rules contained in the health and safety plan (RP)
- Ensure only authorised persons are allowed into any premises where construction work is being undertaken
- Ensure any particulars required to be given in any notice under regulation 7 are displayed in readable condition in a suitable position
- Provide the planning supervisor with appropriate information
- Give directions to any contractor to enable the principal contractor to comply with his duties
- Include in the construction phase health and safety plan 'rules for the management of the construction work'
- Provide comprehensible information to contractors on the risks arising from the construction work
- Ensure every contractor who is an employer provides his employees with appropriate information and suitable training (RP)
- Seek advice from, and views of, persons at work
- Take account of health and safety issues when preparing and presenting tenders.

Duties of contractors

- Co-operate with the principal contractor to enable each of them to comply with the relevant statutory provisions
- Promptly provide the principal contractor with information which might affect the health or safety of any person
- Comply with the directions of the principal contractor
- Comply with any rules applicable to him in the health and safety plan
- Promptly provide the principal contractor with information in relation to any injury, death, condition or dangerous occurrence which the contractor is required to notify under the Reporting of Injuries, Diseases and Dangerous Occurrences Regulations

- Provide the principal contractor with information which needs to be supplied to the planning supervisor.

Duties of employers

- Not to cause to be prepared a design in respect of a project unless he has taken all reasonable steps to ensure that the client for that project is aware of the duties to which the client is subject by virtue of these regulations and practical guidance from the HSC
- Not to cause any or permit any employee to work on construction work unless that person has been provided specific information.

Duties of persons

- Ensure designer is competent to prepare the design
- Ensure contractor has the competence to carry out, or, as the case may be, manage that construction work
- Prior to arranging for a contractor to carry out or manage construction work, ensure he is reasonably satisfied that the contractor has allocated, or will allocate, adequate resources to enable the contractor to comply with the requirements and prohibitions imposed upon him by or under the relevant statutory provisions
- Prior to appointment of a planning supervisor, ensure client is reasonably satisfied that the person he intends to appoint has allocated, or will allocate, adequate resources to enable him to perform the functions of planning supervisor.

☞ **1(d) Approved codes of practice**
   Managing construction for health and safety
☞ **1(e) HSE guidance notes**
   Managing contractors
☞ **3(a) Tables and figures**
   Construction (Design and Management) Regulations 1994: How to decide when the exceptions to the CDM Regulations apply
☞ **3(b) Forms**
   Construction (Design and Management) Regulations 1994: Notification of Project (Form 10)

Construction (Head Protection) Regulations 1989

Responsible for enforcement
    HSE
Level of duty
    Absolute
Duties of employers
- Provide suitable head protection for their employees
- Ensure maintenance of head protection and replacement of parts as necessary
- Ensure head protection is worn
- Make rules or directions as to the wearing of head protection in specified circumstances.

Duties of employees
- Report loss of, or defect in, head protection to employer.

 1(a) Legal background
    Contractors
 1(e) HSE guidance notes
    A guide to the Construction (Head Protection) Regulations 1989

Construction (Health, Safety and Welfare) Regulations 1996

Responsible for enforcement
    HSE
Level of duty
    Absolute
Duties of employers
- Ensure a safe place of work and safe means of access to and from that place of work
- Prevent falls from heights by physical precautions or use of fall arrest equipment

- Provide and maintain physical precautions to prevent falls through fragile materials
- Ensure erection of scaffolds, access equipment, harnesses and nets is under the supervision of a competent person
- Ensure safe use of ladders
- Take steps to prevent materials or objects falling
- Take precautions to prevent people from being struck by falling objects
- Prohibit the throwing of any materials or objects from a height if they could strike someone
- Store materials and equipment safely
- Prevent accidental collapse of new or existing structures or those under construction
- Ensure any dismantling or demolition is planned and carried out in a safe manner under the supervision of a competent person
- Only fire explosive charges after ensuring no one is exposed to risk or injury
- Prevent the collapse of ground both in and above excavations
- Identify and prevent risk from underground cables and other services
- Ensure cofferdams and caissons are properly designed, constructed and maintained
- Take steps to avoid people falling into water or other liquid (RP)
- Ensure PPE and rescue equipment is immediately available for use and maintained in the event of a fall
- Ensure transport by water is under the control of a competent person
- Ensure construction sites are organised so that pedestrians and vehicles can move safely
- Ensure routes are suitable and sufficient for people or vehicles using them
- Prevent or control the movement of vehicles
- Ensure arrangements for giving warnings of any possible dangerous movements of vehicles
- Ensure safe operation of vehicles

- Ensure doors and gates which could present danger are provided with suitable safeguards
- Prevent risk from fire, explosion, flooding and asphyxiation
- Provide emergency routes and exits
- Provide arrangements for dealing with emergencies, including evacuation procedures
- Where necessary, provide fire-fighting equipment, fire detectors and alarm systems
- Provide sanitary and washing facilities and an adequate supply of drinking water
- Provide rest facilities and facilities to change and store clothing
- Ensure sufficient fresh or purified air is available at every workplace and that associated plant is capable of giving visible or audible warning of failure
- Ensure a reasonable working temperature is maintained in indoor workplaces
- Provide facilities for protection against adverse weather conditions
- Ensure suitable and sufficient emergency lighting
- Ensure suitable and sufficient lighting is available, including secondary lighting where appropriate
- Maintain sites in good order and in a reasonable state of cleanliness
- Ensure perimeter of a site is marked by suitable signs
- Ensure all plant and equipment is safe, of sound construction and used and maintained so that it remains safe
- Ensure construction activities where training, technical knowledge or experience is necessary to reduce risks are only carried out by people who meet these requirements or, if not, are supervised by those with appropriate training, knowledge or experience
- Before work at height, on excavations, cofferdams or caissons begins, ensuring the place of work is inspected and, at subsequent specified periods, by a competent person
- Following inspections, ensuring written reports are made by the competent person.

1(a) Legal background
    Contractors
 1(d) Approved codes of practice
    Managing construction for health and safety
 1(e) HSE guidance notes
    Avoiding danger from underground services
    Backs for the future: safe manual handling in construction
    Electrical safety on construction sites
    Fire safety in construction: guidance for clients, designers and those managing and carrying out construction work involving significant risks
    Health and safety in construction
    Health and safety in excavations
    Health and safety in roof work
    Maintaining portable and transportable electrical equipment
    The safe use of vehicles on construction sites
    Work with asbestos cement
 2(b) Hazard checklists
    Construction activities
 3(a) Tables and figures
    Demolition methods
    Places of work requiring inspection by a competent person under Regulation 29(1) of the Construction (Health, Safety and Welfare) Regulations 1996

Control of Asbestos at Work Regulations 2002

Responsible for enforcement
    HSE and local authorities
Level of duty
    Absolute
Defence
    All reasonable precautions and all due diligence
Duties of duty holders
- Undertake suitable and sufficient assessment of asbestos risks

- Review assessment if no longer valid or significant change in the premises
- Record conclusions of the assessment
- Where asbestos is, or is liable to be, present, determine risk from asbestos, prepare written plan indicating parts of premises concerned and measures for managing the risk; plan to be reviewed and revised at regular intervals.

Duties of employers

- Before commencing work liable to create asbestos exposure, identify by analysis or otherwise the type of asbestos involved or, assuming it is not chrysotile alone, has treated it accordingly
- Undertake a suitable and sufficient assessment of asbestos risks
- Record significant findings of the assessment
- Implement steps to meet requirements of regulations
- Prepare suitable written plan of work before commencing work
- Notify enforcing authority of particulars specified in Schedule 1 at least 14 days before commencing work
- Provide information, instruction and training to employees
- Prevent exposure of employees to asbestos (RP)
- Where not RP to prevent, reduce exposure to lowest level RP by measures other than the use of RPE and ensure number of employees exposed is as low as is RP
- In manufacturing processes, substitute for asbestos a safe substance or one of lesser risk (P)
- Where not P to substitute ensure, firstly, adequate design of processes, systems and engineering controls and use of suitable equipment and materials and, secondly, control of exposure at source, including adequate ventilation systems and organisational measures
- Where not RP to reduce exposure to below control limits then, in addition, provide RPE
- Take immediate steps to remedy situation where control limit is exceeded

- Where used in, or produced by, a work process, quantity of asbestos and materials containing asbestos, to be reduced to as low a level as is RP
- Ensure control measures are used or applied (RP)
- Maintain control measures in an efficient state, in efficient working order and in good repair
- Ensure thorough examination and test of LEV systems and RPE at suitable intervals by a competent person
- Maintain record of above examinations and tests
- Provide adequate and suitable protective clothing for employees
- Ensure installation of procedures for dealing with an accident, incident or emergency related to the use, removal or repair of asbestos, provide information on emergency arrangements and ensure warning and other communication systems are established
- Designate asbestos areas and respirator zones, clearly and separately demarcated and identified by notices; control access; prohibit eating, drinking and smoking in such areas
- Monitor exposure of employees to asbestos; maintain records
- Ensure compliance with criteria for air testing and analysis of samples
- Maintain personal health records; keep available for at least 40 years
- Ensure employees exposed to asbestos are under adequate medical surveillance; provide facilities for medical surveillance
- Ensure suitable person informs employees of identifiable disease or adverse health effect as a result of exposure
- Provide adequate washing and changing facilities
- Ensure that raw asbestos or waste containing asbestos is not stored, received or distributed within a workplace unless in a clearly marked sealed container.

Duties of suppliers

- Ensure suitable labelling of products for use at work containing asbestos.

Duties of employees
- Make full and proper use of any control measure, return RPE to accommodation after use and report defects in RPE
- Present themselves for medical examination and tests, provide information to relevant doctor.

 1(d) Approved codes of practice
Control of asbestos at work
The management of asbestos in non-domestic premises
Work with asbestos insulation, asbestos coating and asbestos insulation board
Work with asbestos that does not normally require a licence

 1(e) HSE guidance notes
A comprehensive guide to managing asbestos in premises
Asbestos essentials task manual
Introduction to asbestos essentials
Work with asbestos cement

Control of Lead at Work Regulations 2002

Responsible for enforcement
HSE
Level of duty
Absolute
Duties of employers
- Not to use a glaze other than a leadless glaze or low solubility glaze in the manufacture of pottery
- Not to employ a young person or woman of reproductive capacity in any activity specified in Schedule 1
- Not to carry out work which is liable to expose any employees to lead unless he has made a suitable and sufficient assessment of the risks to the health of those employees and of the steps that need to be taken to meet the requirements of the Regulations
- Review any risk assessment if no longer valid and in specified circumstances

- Prevent, or where this is not reasonably practicable, adequately control exposure of employees to lead
- Avoid the use of lead by replacing it with a substance or process which eliminates or reduces the risk to health (RP)
- Provide employees with suitable and sufficient protective clothing
- Control of exposure, so far as the inhalation of lead is concerned, shall only be treated as being adequate if the occupational exposure limit is not exceeded or, where it is exceeded, the employer identifies the reasons and takes immediate steps to remedy the situation
- PPE to comply with the Personal Protective Equipment Regulations and, in the case of RPE, be of an approved type or conform to a standard approved by the HSE
- Take all reasonable steps to ensure any control measure, other thing or facility is properly used or applied
- Ensure employees do not eat, drink or smoke in any place contaminated, or liable to be contaminated, by lead (RP)
- Maintain plant and equipment, including engineering controls and personal protective equipment, in an efficient state, in efficient working order, in good repair and in a clean condition
- Ensure systems of work and supervision and any other measures are reviewed at suitable intervals and revised if necessary
- Where engineering controls are provided, ensure thorough examination and testing is carried out:
  - in the case of local exhaust ventilation plant, at least once every 14 months
  - in any other case, at suitable intervals
- Where RPE is provided, ensure that thorough examination and tests are carried out at suitable intervals
- Keep suitable records of the above examinations and tests for at least 5 years from the date on which it was made
- PPE to be properly stored, checked at suitable intervals, repaired or replaced as necessary
- PPE which is contaminated to be kept apart from uncontaminated PPE

- Where there is risk of significant exposure to lead, ensure any concentration in air to which employees are exposed is measured in accordance with a suitable procedure at least every 3 months
- Ensure a suitable record of monitoring is maintained
- Provide medical surveillance for each employee who is or is liable to be exposed to lead
- Maintain health record of exposed employees for 40 years after the date of the last entry
- Provide employees with suitable and sufficient information, instruction and training
- Provide specific information and training
- Ensure arrangements to deal with accidents, incidents and emergencies.

Duties of employees
- Make full and proper use of any control measure, other thing or facility provided and, where relevant,
    - take all reasonable steps to ensure it is returned after use to any accommodation provided for it and
    - if he discovers a defect therein, report it forthwith to his employer
- Not eat, drink or smoke in any place which he has reason to believe is contaminated by lead.

 1(d) Approved codes of practice
Control of lead at work

Control of Major Accident Hazards Regulations 1999

Responsible for enforcement
HSE and local authorities
Level of duty
Absolute

Duties of operators

- Take all measures necessary to prevent major accidents and limit their consequences to persons and the environment
- Prepare and keep a document setting out policy with respect to the prevention of major accidents (a 'major accident prevention policy document')
- Within a reasonable period of time prior to the start of construction of an establishment, send a notification to the competent authority containing the information specified in Schedule 3
- Notify the competent authority in the event of:
  - any significant increase in the quantity of dangerous substances notified
  - the nature of physical form, the processes employing them
  - any other information notified to the competent authority in respect of the establishment
- Within a reasonable period of time prior to the start of construction, send a safety report to the competent authority
- Review the safety report:
  - at least every 5 years
  - whenever necessary because of new facts or to take account of new technical knowledge about safety matters
  - whenever a change to the safety management system has been made which could have significant repercussions with respect to the prevention of major accidents or the limitation of consequences of major accidents to persons and the environment
- Prepare an emergency plan (an 'on-site emergency plan')
- Supply information to the local authority and any additional information to enable the off-site emergency plan to be prepared.

Duties of persons who have prepared emergency plans

- Shall at intervals not exceeding 3 years:
  - review and revise the plan
  - test the plan and take reasonable steps to arrange for the emergency services to participate in the test

- Any such review shall take into account changes occurring in the establishment to which the plan relates and within the emergency services concerned, new technical knowledge, and knowledge concerning the response to major accidents
- Take reasonable steps to put the emergency plan into effect without delay when:
    - a major accident occurs or
    - an uncontrolled event occurs which could reasonably be expected to lead to a major accident
- Provide information to members of the public who are in an area liable to be affected by a major accident which area has been notified to the operator by the competent authority
- Consult the local authority with regard to information to be provided to the above members of the public.

Duties of local authority
- Prepare an emergency plan (an 'off-site emergency plan') in respect of the establishment
- Consult the operator, the competent authority, the emergency services, each health authority for the area in the vicinity of the establishment and members of the public
- May charge the operator a fee for performing its functions.

Duties of competent authority
- In view of information contained in a safety report exempt a local authority from the requirement to prepare an off-site emergency plan.

 2(a) Health and safety in practice
   Major incidents

Control of Noise at Work Regulations 2005

Responsible for enforcement
   HSE and local authorities

Level of duty

Absolute and reasonably practicable

Defence

None

Duties of employers

- Where work is liable to expose employees to noise at or above a lower exposure action value, an employer shall make a suitable and sufficient assessment of the risk from the noise to the health and safety of those employees
- The risk assessment shall identify the measures which need to be taken to comply with these Regulations
- Assessment of noise shall be by means of:
  - (a) observation of specific work practices
  - (b) reference to relevant information on probable levels of noise corresponding to any equipment used and
  - (c) if necessary, measurement of the level of noise to which his employees are likely to be exposed
- Assessment shall identify whether any employees are likely to be exposed to noise at or above;
  - (a) a lower exposure action value (LEAV) i.e.
    - (i) a daily or weekly personal noise exposure of 80 dB (A-weighted) or
    - (ii) a peak sound pressure of 135 dB (C-weighted)
  - (b) an upper exposure action value (UEAV), i.e.
    - (i) a daily or weekly personal noise exposure of 85 dB (A-weighted) and
    - (ii) a peak sound pressure of 137 dB (C-weighted) or
  - (c) an exposure limit value (ELV), i.e.
    - (i) a daily or weekly personal noise exposure of 87 dB (A-weighted) and
    - (ii) a peak sound pressure of 140 dB (C-weighted)
- Risk assessment shall include consideration of:
  - (a) the level, type and duration of exposure, including any exposure to peak sound pressure
  - (b) effects of exposure on employees or groups of employees whose health is at particular risk from such exposure
  - (c) any effects on the health and safety of employees resulting from the interaction between noise and the

use of ototoxic substances at work, or between noise and vibration (P)

(d) any indirect effects on the health and safety of employees resulting from the interaction between noise and audible warning signals or other sounds that need to be audible in order to reduce risk at work

(e) any information provided by manufacturers of work equipment

(f) the availability of alternative equipment designed to reduce the emission of noise

(g) any extension of exposure to noise at the workplace beyond normal working hours, including exposure in rest facilities supervised by the employer

(h) appropriate information obtained following health surveillance and

(i) the availability of personal hearing protectors with adequate attenuation characteristics

- Review risk assessment regularly and forthwith if:
  (a) there is reason to suspect it is no longer valid or
  (b) there has been a significant change in the work to which it relates, and where, as a result of review, changes to the risk assessment are required, those changes shall be made

- Consult the employees concerned or their representatives on the assessment of risk.

- Record:
  (a) significant findings of the risk assessment and
  (b) the measures he has taken and which he intends to take to meet requirements of regulations 6 (elimination or control of exposure), 7 (hearing protection) and 10 (information, instruction and training

- Ensure risk is either eliminated at source or, where not RP, reduced to as low a level as is RP

- If an employee is likely to be exposed to noise at or above a UEAV, reduce exposure to as low a level as is RP by establishing and implementing a programme of organisational and technical measures, excluding the provision of personal hearing protectors, which is appropriate to the activity

- Ensure the action taken in accordance with the above is based on the general principles of prevention set out in Schedule 1 to the MHSWR 1999 and shall include consideration of:
  (a) other working methods which reduce exposure to noise
  (b) choice of appropriate work equipment emitting the least possible noise, taking account of the work to be done
  (c) the design and layout of workplaces, work stations and rest facilities
  (d) suitable and sufficient information and training for employees, such that work equipment may be used correctly, in order to minimise exposure to noise
  (e) reduction of noise by technical means
  (f) appropriate maintenance programmes for work equipment, the workplace and workplace systems
  (g) limitation of the duration and intensity of exposure to noise and
  (h) appropriate work schedules with adequate rest periods
- Shall:
  (a) ensure employees are not exposed to noise above an ELV or
  (b) if an ELV is exceeded forthwith:
      (i) reduce exposure to noise to below the ELV
      (ii) identify the reason for that ELV being exceeded and
      (iii) modify the organisational and technical measures taken to prevent it being exceeded again
- Ensure, where rest facilities are made available to employees, exposure to noise in these facilities is reduced to a level suitable for their purpose and conditions of use
- Adapt any measure taken to take account of any employee or group of employees whose health is likely to be particularly at risk from exposure to noise
- Consult employees or their representatives on measures taken to eliminate or control exposure to noise
- Make personal hearing protectors (PHP) available upon request to any employee who carries out work which is likely to exposure him to noise at or above a LEAV

- Provide PHP to any employee so exposed where unable by other means to reduce the levels of noise to below a UEAV
- Ensure that in any area where an employee is likely to be exposed to noise at or above a UEAV:
  (a) the area is designated a Hearing Protection Zone (HPZ)
  (b) the area is demarcated and identified by means of the appropriate sign
  (c) access is restricted where this is practicable and the risk from exposure justifies it
  (d) no employee enters a HPZ unless wearing PHP
- Any PHP made available or provided shall be selected:
  (a) so as to eliminate the risk to hearing or reduce the risk (SFRP) and
  (b) after consultation with employees or their representatives
- Ensure anything provided in compliance with duties under these Regulations is:
  (a) fully and properly used (P)
  (b) maintained in an efficient state, in efficient working order and in good repair
- Where risk assessment identifies hearing risk, ensure such employees are placed under suitable health surveillance, which shall include testing of hearing
- Make and maintain health record of above employees
- Allow employees access to personal health records
- Provide enforcement authority with copies of health records
- Where an employee has identifiable hearing damage, ensure:
  (a) examination of employee by a doctor and/or specialist
  (b) a suitably qualified person informs the employee accordingly
  (c) the risk assessment is reviewed
  (d) existing protective measures are reviewed
  (e) consideration is given to assigning the employee to alternative work where there is no risk from further exposure, taking into account advice from a doctor or occupational health professional and

- (f) continued health surveillance and provide for a review of any other employee who has been similarly exposed
- Provide suitable and sufficient information, instruction and training where employees are exposed to noise likely to be at or above the LEAV
- Information, instruction and training provided shall include:
  - (a) nature of risks from exposure to noise
  - (b) organisational and technical measures to eliminate or control exposure
  - (c) ELVs, UEAVs and LEAVs
  - (d) significant findings of risk assessments, including any measurements taken
  - (e) availability and provision of PHP and their correct use
  - (f) why and how to detect and report signs of hearing damage
  - (g) entitlement to health surveillance
  - (h) safe working practices to minimise exposure to noise and
  - (i) the collective results of any health surveillance
- Information, instruction and training to be updated to take account of significant changes in the type of work and working methods
- Ensure any person, whether or not his employee, has suitable and sufficient information, instruction and training.

Duties of employees
- Make full and proper use of PHP and other control measures provided by employer
- Report any defect in PHP or other control measures
- Present themselves for health surveillance procedures as may be required where found to have identifiable hearing damage.

☞ **1(e) HSE guidance notes**
   Reducing noise at work
   Sound solutions: techniques to reduce noise at work
☞ **2(b) Hazard checklists**
   Noise

3(a) Tables and figures
Decibels (addition of)
Noise control methods
Octave bands (standard range)

Control of Substances Hazardous to Health Regulations 2002

Responsible for enforcement
HSE and local authorities
Level of duty
So far as is reasonably practicable
Defence
All reasonable precautions and all due diligence
Duties of employers

- Not carry out work which is liable to expose employees to a substance hazardous to health (SHH) unless he has made a suitable and sufficient assessment of the risks and the steps that need to be taken, and implemented these steps
- The risk assessment shall be reviewed regularly and forthwith if no longer valid, if there is a significant change in the work and results of monitoring show it to be necessary
- Where changes to the risk assessment are required, these changes shall be made
- Where 5 or more employees, significant findings and steps to prevent or control exposure to be recorded
- Ensure exposure of employees is either prevented or adequately controlled
- Substitution to be undertaken to avoid use of a SHH
- Where not RP to prevent exposure, exposure must be controlled by, in order of priority, design and use of appropriate work processes, engineering controls and suitable work equipment and materials, control of exposure at source, including appropriate ventilation systems and organisational measures, and the provision of suitable PPE in addition to the above measures

- Where not RP to prevent exposure to a carcinogen or a biological agent, specific control measures must be applied
- Where there is exposure to a SHH, control shall only be treated as adequate if:
  - (a) the principles of good practice for the control of exposure to SHH set out in Schedule 2A to the COSHH (Amendment) Regulations 2004 are applied
  - (b) any workplace exposure limit for that substance is not exceeded
  - (c) for a substance
    - (i) which carries the risk phrase R45, R46 or R49, or for a substance listed in Schedule 1
    - (ii) which carries the risk phrase R42 or R42/43, or which is listed in Section C of HSE publication 'Asthmagen: Critical assessments of the evidence for agents implicated in occupational asthma' as updated from time to time, or any other substance which the risk assessment has shown to be a potential cause of occupational asthma exposure is reduced to as low a level as is RP.
- Take all reasonable steps to ensure control measures are used by employees
- Maintain plant and equipment, including engineering controls and PPE, in an efficient state, in efficient working order, in good repair and in a clean condition
- Ensure review of provision of systems of work and supervision and any other measures at suitable intervals
- Ensure thorough examination and testing of LEV systems at least once every 14 months, and of RPE at suitable intervals
- Keep a suitable record of above examinations and tests
- Ensure PPE is properly stored, checked at suitable intervals and repaired or replaced when defective
- Ensure exposure to SHH is monitored in accordance with a suitable procedure
- Ensure employees are provided with suitable health surveillance, particularly where exposed to a specified substance and related process

- Maintain an appropriate personal health record; keep available for 40 years
- Provide information, instruction and training for persons exposed to SHH
- Ensure provision of arrangements to deal with accidents, incidents and emergencies
- In the case of certain fumigants, that is, hydrogen cyanide, phosphine or methyl bromide, take specified precautions.

Duties of employees
- Make full and proper use of any control measure, other thing or facility provided
- Take all reasonable steps to ensure it is returned after use to the accommodation provided
- Report defects in the above to his employer.

☞ **1(d) Approved codes of practice**
Control of substances hazardous to health
Control of substances hazardous to health in fumigation operations

☞ **1(e) HSE guidance notes**
An introduction to local exhaust ventilation
Assessing and managing risks at work from skin exposure to chemical agents
A step-by-step guide to COSHH assessment
Biological monitoring in the workplace
Choice of skin care products for the workplace
Control of legionellosis, including legionnaire's disease
Control of substances hazardous to health in fumigation operations
COSHH essentials – easy steps to control chemicals
Cost and effectiveness of chemical protective gloves for the workplace
Dust: general principles of prevent ion
Health risk management: A guide to working with solvents
Health surveillance at work
Introduction to local exhaust ventilation
Maintenance, examination and testing of local exhaust ventilation

Monitoring strategies for toxic substances
Preventing asthma at work
*The election, use and maintenance of respiratory protective
    equipment*
Workplace exposure limits

 2(b) Hazard checklists
Hazardous substances

 3(a) Tables and figures
Categories of danger
*Hazardous substances that can be revealed by medical
    analysis*
Local exhaust ventilation systems

 3(b) Forms
Health risk assessment

Control of Vibration at Work Regulations 2005

Responsible for enforcement
   Health and Safety Executive or local authority
Level of duty
   Absolute
Defence
   None
Duties of employers
   • Make a suitable and sufficient assessment of the risks to
     employees created by work where they may be liable to
     be exposed to vibration; the risk assessment shall identify
     the measures that need to be taken to comply with the
     regulations
   • When undertaking risk assessment, assess daily exposure
     by means of:
     (a) observation of specific working practices
     (b) reference to relevant information on the probable mag-
         nitude of the vibration corresponding to the equipment
         used in the particular working conditions

(c) if necessary, measurement of the magnitude of vibration to which his employees are likely to be exposed and the employer shall assess whether any employees are likely to be exposed above the actual exposure action value or an exposure limit value.

Exposure limit values and action values

- For hand-arm vibration:
  - (a) the daily exposure limit value normalised to an 8-hour reference period is 5 m/s^2
  - (b) the daily exposure action value normalised to an 8-hour reference period is 2.5 m/s^2
  - (c) daily exposure shall be ascertained on the basis set out in Schedule 1 Part 1
- For whole body vibration:
  - (a) the daily exposure limit value normalised to an 8-hour reference period is 1.15 m/s^2
  - (b) the daily exposure action value normalised to an 8-hour reference period is 0.5 m/s^2
  - (c) daily exposure shall be ascertained on the basis set out in Schedule 2 Part 1
- The risk assessment shall include consideration of:
  - (a) the magnitude, type and duration of exposure, including any exposure to intermittent vibration or repeated shocks
  - (b) the effects of exposure to vibration on employees whose health is at particular risk from such exposure
  - (c) any effects of vibration on the workplace and work equipment, including the proper handling of controls, the reading of indicators, the stability of structures and the security of joints
  - (d) any information provided by the manufacturers of work equipment
  - (e) the availability of replacement equipment designed to reduce exposure to vibration
  - (f) any extension of exposure at the workplace to whole-body vibration beyond normal working hours, including exposure in rest facilities supervised by the employer

- (g) specific working conditions such as low temperatures
- (h) appropriate information obtained from health surveillance including, where possible, published information
- The risk assessment shall be reviewed regularly and forthwith if there is reason to suspect it is no longer valid or there has been a significant change in the work to which it relates
- Employer shall record the significant findings of the risk assessment and the measures which he has taken or intends to take to meet requirements relating to elimination or control of exposure to vibration
- On the basis of the general principles of prevention set out in Schedule 1 of the MHSWR, the employer shall ensure risk is either eliminated at source or reduced (RP)
- Where it is not RP to eliminate risk at source and the risk assessment indicates that an exposure action value is likely to be exceeded, the employer shall reduce exposure as low as is RP by establishing and implementing a programme or organisational and technical measures which is appropriate to the activity and consistent with the risk assessment, and shall include consideration of:
  - (a) other methods of work which eliminate or reduce exposure
  - (b) choice of work equipment of appropriate ergonomic design which produces the least possible vibration
  - (c) provision of auxiliary equipment which reduces the risk of injuries caused by vibration
  - (d) appropriate maintenance programmes for work equipment, the workplace and workplace systems
  - (e) the design and layout of workplaces, work stations and rest facilities
  - (f) suitable and sufficient information and training for employees
  - (g) limitation of the duration and intensity of exposure
  - (h) appropriate work schedules with adequate rest periods and
  - (i) the provision of clothing to protect employees from cold and damp

- The employer shall:
  - (a) ensure that his employees are not exposed to vibration above an exposure limit value or
  - (b) if an exposure limit value is exceeded:
    - (i) take action to reduce exposure to below the limit value;
    - (ii) identify the reason for the limit being exceeded; and
    - (iii) modify the organisational and technical measures taken
- The above paragraph shall not apply where exposure of an employee is usually below the exposure action value but varies markedly from time to time and may occasionally exceed the exposure limit value, provided that:
  - (a) any exposure to vibration averaged over one week is less than the exposure limit value
  - (b) there is evidence to show that the risk from the actual pattern of exposure is less than the corresponding risk from constant exposure at the exposure limit value
  - (c) risk is reduced as low as RP, taking into account the special circumstances
  - (d) the employees concerned are subject to increased health surveillance where such surveillance is appropriate and exposure within the meaning of this paragraph shall be ascertained on the basis set out in Schedule 1 Part II for hand-arm vibration and Schedule 2 Part II for whole-body vibration
- The employer shall adapt any measure taken to take account of any employee who is particularly sensitive to vibration.
- If:
  - (a) the risk assessment indicates there is a risk to employees exposed to vibration or
  - (b) employees are exposed to vibration in excess of an exposure action value
  the employer shall ensure that such employees are under suitable health surveillance.

- Health surveillance shall be appropriate where the exposure of the employee is such that
  - (a) a link can be established between that exposure and an identifiable disease or adverse health effect
  - (b) it is probable that the disease or effect may occur under particular conditions of work and
  - (c) there are valid techniques for detecting the disease or effect
- The employer shall ensure that a health record in respect of each of his employees who undergoes health surveillance is made and maintained and that record or copy of same is kept available in a suitable form

  The employer shall:
  - (a) on reasonable notice, allow an employee access to his health record
  - (b) provide the enforcing authority with copies of records and
  - (c) if he ceases to trade, notify the HSE forthwith and make available to the HSE any records kept by him
- Where any employee is found to have an identifiable disease or adverse health effect which is considered by a doctor or other occupational health professional to be the result of exposure to vibration, the employer shall:
  - (a) ensure that a suitably qualified person informs the employee and provides information about further health surveillance
  - (b) ensure that he is informed himself of any significant findings from the employee's health surveillance
  - (c) review the risk assessment
  - (d) review any elimination or control measure
  - (e) consider assigning the employee to alternative work where there is no risk from further exposure and
  - (f) provide for a review of the health of any other employee who has been similarly exposed, including a medical examination
- If:
  - (a) the risk assessment indicates there is a risk of exposure to vibration or

    (b) employees are exposed to vibration in excess of an exposure action value

the employer shall provide those employees and their representatives with suitable and sufficient information, instruction and training

- The information, instruction and training shall include:
  - (a) the organisational and technical measures taken
  - (b) the exposure limit values and action values
  - (c) the significant findings of the risk assessment, including any measurements taken
  - (d) why and how to detect and report signs of injury
  - (e) entitlement to appropriate health surveillance
  - (f) safe working practices to minimise exposure to vibration and
  - (g) the collective results of any health surveillance undertaken
- The information, instruction and training shall be adapted to take account of significant changes in the type of work carried out or methods of work used by the employer.

Duties of employees

- When required by the employer and at the cost of his employer, present himself during working hours for such health surveillance as may be required.

 1(e) HSE guidance notes
Hand-arm vibration
Vibration solutions

Dangerous Substances and Explosive Atmospheres Regulations 2002

Responsible for enforcement
    Health and Safety Executive
Level of duty
    Absolute
Duties of employers

- Make a suitable and sufficient assessment of the risks to employees arising from a dangerous substance:
  - eliminate or reduce the risk (RP) or
  - substitute with a substance or process which either eliminates or reduces risk or
  - apply specified measures to control the risk and to mitigate detrimental effects of fire or explosion or other harmful physical effects (RP)
- Classify places at the workplace where explosive atmospheres may occur and designate as hazardous zones
- Ensure safety of employees from accident, incident or emergency through:
  - (a) procedures including first aid facilities and relevant safety drills
  - (b) provision of information on emergency arrangements
  - (c) suitable warning and other communication systems
  - (d) before explosion conditions are reached, installation of visual or audible warnings
  - (e) provide and maintain escape facilities where risk assessment indicates such provision is necessary
- In the event of accident, incident or emergency:
  - (a) ensure immediate steps are taken to remedy situation
  - (b) regulate access to affected area, providing appropriate PPE and specialised safety equipment and plant
- Provide information, instruction and training for employees
- Identify hazardous contents of containers and pipes
- Co-ordinate implementation of all measures required with other employers who share the workplace.

Schedules

1. General safety measures – workplace and work processes, organisational measures
2. Classification of places where explosive atmospheres may occur
3. Criteria for the selection of equipment and protective systems
4. Warning sign for places where explosive atmospheres may occur
5. Legislation concerned with the marking of containers and pipes.

1(d) Approved codes of practice
Dangerous Substances and Explosives Atmospheres Regulations
Unloading petrol from road tankers
Design of plant, equipment and workplaces
Storage of dangerous substances
Control and mitigation measures
Safe maintenance, repair and cleaning procedures
Dangerous substances and explosive atmospheres

 1(e) HSE guidance notes
The safe use and handling of flammable liquids
The storage of flammable liquids in containers
The storage of flammable liquids in tanks

 2(b) Hazard checklists
Flammable substances
Hazardous substances

Electricity at Work Regulations 1989

Responsible for enforcement
   HSE and local authorities
Level of duty
   Reasonably practicable
Defence
   All reasonable steps and all due diligence
Duties of employers, self-employed persons and managers of mines and quarries

- All systems at all times to be constructed and maintained so as to prevent danger
- Work activities to be carried out in such a manner as not to give rise to danger
- Equipment for protecting persons at work on or near electrical equipment to be suitably maintained in a condition suitable for that use and properly used
- No electrical equipment to be put into use where its strength and capability may be exceeded in such a way as to give rise to danger

- Equipment used in hazardous environments to be of such construction or protected as to prevent danger from such exposure
- All conductors which may give rise to danger to be suitably covered with insulating material and protected, or have precautions taken in respect of them as will prevent danger
- Precautions to be taken, either by earthing or other suitable means, to prevent danger arising when any conductor (other than a circuit conductor) which may reasonably foreseeably become charged as a result of either the use of the system, or a fault in a system
- Suitable precautions to be taken to ensure integrity of referenced conductors
- Every joint and connection in a system to be mechanically and electrically suitable for use
- Efficient means, suitably located, to be provided for protecting from excess of current
- Suitable means to be provided for cutting off the supply of electrical energy and for isolation of any electrical equipment
- Adequate precautions to be taken for work on equipment made dead
- Suitable precautions to be taken where persons are engaged on work on or near live conductors
- Adequate working space, access and lighting to be provided at work in circumstances which may give rise to danger
- Persons engaged in any work activity where technical knowledge or experience is necessary to prevent danger or risk of injury to be competent or under such degree of supervision as may be appropriate
- Part III of the Regulations applies to mines only.

☞ **1(e) HSE guidance notes**

Electrical safety in arc welding
Electrical safety on construction sites
Electricity at work: safe working practices
Keeping electrical switchgear safe

Maintaining portable and transportable electrical equipment
Memorandum of Guidance on the Electricity at Work Regulations 1989

 2(b) Hazard checklists
Electrical equipment

Health and Safety (Consultation with Employees) Regulations 1996

Responsible for enforcement
    HSE, local authorities
Level of duty
    Absolute
Duties of employers
- Must consult any employees who are not covered by the Safety Representatives and Safety Committees Regulations 1977
- Consultation may be direct or through appointed representatives of employee safety.

 1(e) HSE guidance notes
A guide to the Health and Safety (Consultation with Employees) Regulations 1996

Health and Safety (Display Screen Equipment) Regulations 1992

Responsible for enforcement
    HSE and local authorities
Level of duty
    Absolute
Duties of employers
- Perform a suitable and sufficient analysis of those workstations used for the purposes of his undertaking by defined 'users' and 'operators'

- Review any risk analysis if no longer valid or significant change in the matters to which it relates
- Reduce risks to the lowest extent (RP): See Hazard Checklists – Display screen equipment
- Ensure any workstation meets the requirements laid down in the Schedule to the Regulations
- Plan activities of users and operators to ensure screen breaks or change of activities
- Provide appropriate eye and eyesight tests for users
- Provide adequate health and safety training in use of workstation
- Provide adequate information on measures taken by employer with respect to analysis of workstation.

 1(e) HSE guidance notes
A pain in your workplace: ergonomic problems and solutions
Display screen equipment work
Seating at work
The law on VDUs
Upper limb disorders in the workplace
Work-related upper limb disorders
Work with display screen equipment

 2(b) Hazard checklists
Display screen equipment

 3(a) Tables and figures
Health and Safety (Display Screen Equipment) Regulations 1992
(a) *Display screen equipment workstation – design and layout*
(b) *Seating and posture for typical office tasks*

Health and Safety (First Aid) Regulations 1981

Responsible for enforcement
HSE, local authorities

Level of duty
   Absolute
Duties of employers
- Make first aid arrangements for all employees, the scale of arrangements being based on the range of work activities and the hazards to which employees are exposed
- Ensure all employees are informed of first aid arrangements.

 1(d) Approved codes of practice
   First aid at work
 **1(e) HSE Guidance**
   The training of first aid at work

Health and Safety (Information for Employees) Regulations 1998

Responsible for enforcement
   HSE, local authorities
Level of duty
   Absolute
Duties of employers
- Must ensure information relating to health, safety and welfare be furnished to employees by means of the poster or leaflet (entitled Health and Safety Law – What you should know)
- Must insert the name and address of the enforcing authority and address of the Employment Medical Advisory Service in the appropriate space on the poster or specified in a written notice accompanying the leaflet.

Health and Safety (Safety Signs and Signals) Regulations 1996

Responsible for enforcement
   HSE, local authorities

Level of duty
   Absolute
Duties of employers
   • Use a safety sign where a risk cannot be adequately avoided or controlled by other means Install, where necessary, road traffic signs
   • Maintain safety signs
   • Explain unfamiliar signs to their employees and tell them what they need to do when they see a safety sign
   • Mark pipework containing dangerous substances
   • Fire exit signs to incorporate the Running Man symbol.

 1(e) HSE guidance
   Safety signs and signals: Health and Safety (Safety Signs and Signals) Regulations 1996: Guidance on Regulations

Highly Flammable Liquids and Liquefied Petroleum Gases Regulations 1972

Responsible for enforcement
   HSE
Level of duty
   Absolute
Duties of occupiers of factories
   • When not in use or being conveyed, all highly flammable liquids (HFLs) (as defined) should be stored in a safe manner
   • Ensure all HFLs are stored in one of the following ways:
      ○ in suitable fixed storage tanks in a safe position
      ○ in suitably closed vessels kept in a safe position in the open air and, where necessary, protected against direct sunlight
      ○ in a suitable closed vessel kept in a store room that is either in a safe position or is of fire-resisting structure
      ○ in the case of a workroom, where the aggregate quantity of HFL stored does not exceed 50 litres, in

suitable closed vessels kept in a suitably placed cupboard or bin that is a fire-resisting structure

- Storage tanks to be provided with a bund wall enclosure that is capable of containing 110% of the capacity of the largest tank within the bund
- The ground beneath storage tanks to be impervious to liquid and be so sloped that any minor spillage will not remain beneath the vessels, but will run away to the sides of the enclosure
- Bulk storage tanks must not be located inside buildings or on the roof of a building
- Underground tanks should not be sited under the floors of process buildings
- A drum storage area should be surrounded by a sill capable of containing the maximum spillage from the largest drum in store
- Every store room, cupboard, bin, tank and vessel used for storing HFLs to be clearly and boldly marked 'Highly Flammable' or 'Flash point below 32°C' or 'Flash Point in the Range 22°C to 32°C'.
- All liquefied petroleum gas (LPG) (as defined) must be stored in one of the following ways:
  - in suitable underground reservoirs or in suitable fixed storage tanks located in a safe position, either underground or in the open air
  - in suitable movable storage tanks or vessels kept in a safe position in the open air
  - in pipelines or pumps forming part of an enclosed system
  - in suitable cylinders kept in a safe position in the open air or, where this is not reasonably practicable, in a store room constructed of non-combustible material, having adequate ventilation, being in a safe position, of fire-resisting structure, and being used solely for the storage of LPG and/or acetylene cylinders
- LPG cylinders must be kept in a store until they are required for use, and any expended cylinder must be returned to the store as soon as is reasonably practicable

- Every tank, cylinder, store room, etc used for the storage of LPG to be clearly and boldly marked 'Highly Flammable – LPG'
- Where HFLs are to be conveyed in a factory, a totally enclosed piped system should be used (RP)
- Where not RP, a system using closed non-spill containers may be acceptable
- Portable vessels, when emptied, should be removed to a safe place without delay
- Where, in any process or operation, any HFL is liable to leak or be spilt, all reasonably practicable steps should be taken to ensure that any such HFL is contained or immediately drained off to a suitable container, or to a safe place, or rendered harmless
- No means likely to ignite vapour from any HFL should be present where there may be dangerous concentrations of vapours from HFL
- Where any HFL is being utilised in a workplace, reasonably practicable steps should be taken so as to minimise the risk of escape of HFL vapours into the general workplace atmosphere: where this cannot be avoided, the safe dispersal of HFL vapours should be effected (RP)
- In cases where explosion pressure relief or adequate natural ventilation are required in a fire-resistant structure, a relaxation of the specification of a fire-resistant structure is allowable
- There must be adequate and safe means of escape in case of fire from every room in which any HFL is manufactured, used or manipulated
- A fire certificate is required where:
  ○ HFLs are manufactured
  ○ LPG is stored
  ○ liquefied flammable gas is stored
- Where, as a result of any process or operation involving any HFL, a deposit of any solid waste residue is liable to give rise to risk of fire on any surface:
  ○ steps must be taken to prevent the occurrence of all such deposits (RP)

- ○ where any such deposits occur, effective steps must be taken to remove all such residues, as often as necessary, to prevent danger
- Appropriate fire-fighting equipment should be made readily available for use in all factories where HFLs are manufactured, used or manipulated.

Duties of all persons

- No person may smoke in any place in which any HFL is present and where the circumstances are such that smoking could give rise to fire.

Duties of employees

- Must comply with the Regulations and co-operate in their implementation
- On discovering any defect in plant, equipment or appliance, report the defect without delay to the occupier, manager or other responsible person.

☞ **1(e) HSE guidance notes**
The safe use and handling of flammable liquids
The storage of flammable liquids in containers
The storage of flammable liquids in tanks

☞ **2(b) Hazard checklists**
Fire safety
Flammable substances

Ionising Radiations Regulations 1999

Responsible for enforcement
HSE
Level of duty
Absolute
Defence
Various (See Regulation 36)
Duties of radiation employers

- Not carry out the following practices, except in accordance with a prior authorisation granted by the HSE:
  - ○ the use of electrical equipment intended to produce X-rays for the purpose of:

- – industrial radiography
- – the processing of products
- – research or
- – the exposure of persons for medical treatment
- ○ the use of accelerators, except electron microscopes
- ○ An authorisation may be subject to conditions
- Give 28 days' notice to the HSE of carrying out work with ionising radiation for the first time
- Prior to commencing a new activity, undertake a risk assessment to identify the measures he needs to take to restrict exposure of employees or other persons to ionising radiation
- Take all necessary steps to restrict exposure (RP)
- Provide personal protective equipment (PPE)
- Maintain and examine engineering controls etc and PPE
- Ensure employees are not exposed to an extent that any dose limit is exceeded in one calendar year
- Where a radiation accident is reasonably foreseeable, prepare a contingency plan
- Consult with radiation protection advisers
- Ensure the provision of:
    - ○ suitable and sufficient information and instruction to radiation workers
    - ○ information to other persons directly concerned with the work
    - ○ information to those female employees with respect to radiation risks to the foetus and to a nursing infant, and of the need to inform the employer of their becoming pregnant or if they are breastfeeding
- Ensure co-operation with other employers where their employees may be exposed
- Designate controlled and supervised areas
- Make and set down in writing local rules in respect of controlled and supervised areas
- Ensure local rules are observed and brought to the attention of appropriate employees
- Appoint one or more radiation protection supervisors to ensure compliance with regulations

- Ensure any designated area is adequately described in local rules
- Ensure retention of monitoring or measurements for two years from the date they were recorded
- Where there is a significant risk of spread of radioactive contamination from a controlled area, make adequate arrangements to restrict such spread (RP)
- Monitor radiation in controlled and supervised areas
- Ensure monitoring equipment is properly maintained and adequately tested and examined
- Maintain records of monitoring and testing
- Designate classified persons
- In the case of classified persons, ensure an assessment of all doses received is made and recorded
- Estimate doses where a dosemeter or device is lost, damaged or destroyed or it is not practicable to assess the dose received by a classified person
- Following an accident or other occurrence where it is likely a classified worker has received an effective dose exceeding 6 mSv or an equivalent dose greater than three-tenths of any relevant dose limit, arrange for a dose assessment to be made by the approved dosimetry service forthwith
- Ensure specified persons are under appropriate medical surveillance
- Arrange for investigation and notification of suspected or actual overexposure
- Ensure a radioactive substance shall be in the form of a sealed source (RP)
- Ensure operation of procedures for accounting for radioactive substances
- Ensure suitable precautions for the moving, transporting or disposing of radioactive substances
- Ensure notification to the HSE of certain occurrences, e.g. spillages
- In the case of equipment used for medical exposure
    - ensure such equipment is of such design or construction and is so installed and maintained as to be

capable of restricting exposure of any person who is undergoing a medical exposure (RP)

○ make arrangements for a suitable quality assurance programme to be provided

○ take steps to prevent the failure of any such equipment

○ where an incident may have occurred as a result of malfunction of, or defect in, radiation equipment, make an immediate investigation of the suspected incident and notify the HSE accordingly

○ following the above investigation:
– in respect of an immediate report, retain same for at least 2 years;
– in respect of a detailed report, retain same for at least 50 years.

Duties of all persons

• Not to enter a controlled area unless he can demonstrate, by personal dose monitoring or other suitable measurements, that the doses are restricted

• Not to intentionally or recklessly misuse or without reasonable excuse interfere with any radioactive substance or any electrical equipment.

Duties of manufacturers etc. of articles for use in work with ionising radiation

• Ensure any article is so designed and constructed as to restrict the extent to which employees and other persons are or are likely to be exposed to radiation

• Where erecting or installing any article for use at work, undertake a critical examination of the way it was erected or installed, consult with radiation protection adviser and the

• radiation employer with information about the proper use, testing and maintenance of the article.

Duties of employees

• Not knowingly expose himself or any other person to ionising radiation to an extent greater than is reasonably necessary

• Exercise reasonable care while carrying out work

- Make full and proper use of any PPE, report defects in PPE to employer, and take steps to ensure PPE is returned to accommodation provided
- Comply with any reasonable requirement imposed on him for the purpose of making measurements and assessments in respect of dose received and dosimetry for accidents
- Present himself for medical examination and tests as may be required
- Notify his employer where he has reasonable cause to believe:
    ○ he or some other person has received an overexposure
    ○ where a release into the atmosphere or spillage has occurred
    ○ radiation incident involving medical exposure has occurred.

Duties of outside workers
- Not to misuse the radiation passbook issued to him or to falsify or attempt to falsify any of the information contained in it.

Schedules
1. Work not required to be notified under Regulation 6
2. Particulars to be provided in a notification under Regulation 6(2)
3. Additional particulars that the Executive may require
4. Dose limits
5. Matters in respect of which a radiation protection adviser must be consulted by a radiation employer
6. Particulars to be entered in a radiation passbook
7. Particulars to be contained in a health record
8. Quantities and concentrations of radionuclides.

☞ **1(d) Approved codes of practice**
   Work with ionising radiation
☞ **1(e) HSE guidance notes**
   Protection of persons against ionising radiation arising from any work activity
☞ **2(a) Health and safety practice**
   Local rules

2(b) Hazard checklists
Radiation hazards
3(a) Tables
Electromagnetic spectrum

Lifting Operations and Lifting Equipment Regulations 1998

Responsible for enforcement
    HSE, local authority
Level of duty
    Absolute
Duties of employers
- Ensure that:
    - lifting equipment is of adequate strength and stability for each load
    - every part of a load and anything attached to it and used in lifting it is of adequate strength
- Ensure safety of lifting equipment for lifting persons
- Ensure lifting equipment is positioned or installed in such a way to prevent specified accidents (RP)
- Ensure there are suitable devices to prevent a person falling down a shaft or hoistway
- Ensure machinery and accessories for lifting loads are clearly marked to indicate their safe working loads (SWLs)
- Ensure where the SWL of machinery for lifting depends upon its configuration:
    - the machinery is clearly marked to indicate its SWL for each configuration
    - information which clearly indicates its SWL for each configuration is kept with the machinery
- Ensure accessories for lifting are clearly marked in such a way that it is possible to identify the characteristics necessary for their safe use
- Ensure lifting equipment which is designed for lifting persons is appropriately and clearly marked to this effect

- Ensure lifting equipment which is not designed for lifting persons, but which might be so used in error, is appropriately and clearly marked to the effect that it is not designed for lifting persons
- Ensure every lifting operation involving lifting equipment is:
  - properly planned by a competent person
  - appropriately supervised
  - carried out in a safe manner
- Ensure, before lifting equipment is put into service for the first time, it is thoroughly examined for any defect unless either:
  - it has not been used before
  - in the case of lifting equipment for which an EC declaration of conformity could or (in the case of a declaration under the Lifts Regulations 1997) should have been drawn up, the employer has received such declaration made not more than 12 months before the lifting equipment is put into service
  - or, if obtained from the undertaking of another person, it is accompanied by physical evidence that the last examination required to be carried out has been carried out
- Ensure, where the safety of lifting equipment may depend upon its installation conditions, it is thoroughly examined:
  - after installation and before being put into service for the firs time
  - after assembly and before being put into service at a new site or in a new location, to ensure it has been installed correctly and is safe to operate.
- Ensure that lifting equipment which is exposed to conditions causing deterioration which is liable to result in dangerous situations is:
  - thoroughly examined:
  - in the case of lifting equipment for lifting persons or an accessory for lifting, at least every six months
  - in the case of other lifting equipment, at least every 12 months, or

- in either case, in accordance with an examination scheme, and each time that exceptional circumstances which are liable to jeopardise the safety of the lifting equipment have occurred
- if appropriate for the purpose, is inspected by a competent person at suitable intervals between thorough examinations
- Ensure that no lifting equipment:
    - leaves his undertaking, or
    - if obtained from the undertaking of another person, is used in his undertaking, unless it is accompanied by physical evidence that the last thorough examination required to be carried out has been carried out
- Where notified of a defect following a thorough examination of lifting equipment, ensure that the lifting equipment is not used:
    - before the defect is remedied, or
    - in specific cases, after a time specified and before the defect is remedied
- Where an employer receives an EC declaration of conformity, he shall keep it for so long as he operates the lifting equipment
- Ensure information contained in:
    - every report is kept available for inspection
    - every record is kept available until the next such record is made.

Duties of a person making a thorough examination of lifting equipment

- Shall:
    - notify the employer forthwith of any defect in the lifting equipment which is, or could become, a danger to persons
    - make a report of the thorough examination in writing authenticated by him or on his behalf by signature or equally secure means and containing the information specified in Schedule 1 to:
        – the employer

- any person for whom the lifting equipment has been hired or leased (P)
  - ○ where there is a defect in the lifting equipment involving an existing or imminent risk of serious personal injury, send a copy of the report to the relevant enforcing authority.

Duties of a person making an inspection of lifting equipment
- Shall:
  - ○ notify the employer forthwith of any defect in the lifting equipment which is, or could become, a danger to persons
  - ○ make a record of his inspection in writing (P).

 1(d) Approved codes of practice
   Safe use of lifting equipment
 1(e) HSE guidance notes
   Safe use of lifting equipment

Lifts Regulations 1997

Responsible for enforcement
   HSE, local authorities
Level of duty
   Absolute
Defence
   All reasonable precautions and all due diligence
   Act or default of another person
   Reliance on information given by another
Duties of responsible persons
- Not to place on the market and put into service any lift unless the following requirements have been complied with:
  - ○ it satisfies the essential health and safety requirements (EHSRs) (as defined) and for the purpose of satisfying those requirements
  - ○ where a transposed harmonised standard covers one or more of the relevant EHSRs, any lift constructed in accordance with that transposed harmonised

    standard shall be presumed to comply with that or, as the case may be, those EHSRs

- ○ by calculation or on the basis of design plans, it is permitted to demonstrate the similarity of a range of equipment to satisfy the essential safety requirements
- ○ the appropriate conformity assessment procedure in respect of the lift has been carried out
- ○ the CE marking has been affixed to it by the installer of the lift
- ○ a declaration of conformity has been drawn up taking account of the specifications given in the Schedule used for the conformity assessment procedure
- ○ it is in fact safe.

- Not to place on the market and put into service any safety component unless the following requirements have been complied with:
  - ○ it satisfies the relevant EHSRs and for the purpose of satisfying those requirements where a transposed harmonised standard covers one or more of the relevant EHSRs, any safety component constructed in accordance with that transposed harmonised standard shall be presumed to be suitable to enable a lift on which it is correctly installed to comply with that or, as the case may be, those EHSRs
  - ○ the appropriate conformity assessment procedure has been carried out
  - ○ the CE marking has been affixed to it, or on a label inseparably attached to it, by the manufacturer or his authorised representative established in the Community
  - ○ a declaration of conformity has been drawn up by the manufacturer or his authorised representative established in the Community containing the information listed in Part A of Schedule 2, taking account of specifications given in the Schedule for the conformity assessment procedure
  - ○ it is in fact safe.

- Retain any technical documentation or information under the conformity assessment procedure for any period specified in that procedure.

Duty of suppliers of lifts or safety components

- Ensure that the lift or safety component is safe.

Duty of persons specified in a conformity assessment procedure

- Retain any technical documentation or other information specified in that respect for any period specified in that procedure.

Duties of a person responsible for work on the building or construction and installer of a lift

- Shall:
    - keep each other informed of the facts necessary for, and
    - take the appropriate steps to ensure the proper operation and safe use of the lift; in particular it shall be ensured that the shafts intended for lifts do not contain any piping or wiring or fittings other than that necessary for the operation and safety of that lift.

Duty of designers of lifts

- The person responsible for the design must supply to the person responsible for construction, installation and testing all necessary documents and information for the latter person to be able to operate in absolute security.

Duty of installers of lifts

- In the case of a lift, supply to the Commission of the European Communities, the Member States and any other notified bodies, on request, a copy of the declaration of conformity, together with a copy of the reports of the tests involved in the final inspection carried out as part of the conformity assessment procedure.

Duty of person who draws up declaration of conformity

- Retain a copy of the declaration of conformity for a period of 10 years:
    - in the case of a lift, from the date on which the lift was placed on the market, and

○ in the case of a safety component, from the date on which safety components of that type were last manufactured by that person.

Duty of notified bodies
- Carry out the procedures and specific tasks for which it has been appointed including (where so provided as part of the procedures) surveillance to ensure that the installer of the lift or manufacturer of the safety component or such other responsible person, as the case may be, duly fulfils the obligations arising out of the relevant conformity assessment procedure.

Management of Health and Safety at Work Regulations 1999

Responsible for enforcement
    HSE and local authorities
Level of duty
    Absolute
Duties of employers
- Make a suitable and sufficient assessment of the risks to the health and safety of his employees and to other persons not in his employment, who may be affected by the activities of his undertaking, for the purposes of identifying the measures he needs to take to comply with the requirements and prohibitions imposed on him by or under the relevant statutory provisions
- Implement preventive and protective measures according to Schedule 1
- Make and give effect to arrangements for the effective planning, organisation, control, monitoring and review of preventive and protective measures
- Where 5 or more employees, record arrangements
- Provide health surveillance where appropriate

- Appoint one or more competent persons to oversee protective measures arising from risk assessment
- Establish procedures for serious and imminent danger and danger areas
- Ensure contacts with external services are arranged particularly with respect to first aid, emergency medical care and rescue work
- Provide information to employees which is comprehensible and relevant information on identified risks, preventive and protective measures, procedures for serious and imminent danger, identities of competent persons and risks associated with shared workplaces
- Provide parents with specific information prior to employing a child
- Ensure co-operation, co-ordination and information on risks to other parties in shared workplaces
- Provide comprehensible information to employers of an outside undertaking on the risks arising from his undertaking and the safety measures necessary
- When entrusting tasks to employees, take into account their capabilities as regards health and safety
- Ensure adequate health and safety training under specified circumstances
- Inform temporary workers of special qualifications or skills required to work safely and of any health surveillance required
- Undertake risk assessment in respect of new or expectant mothers
- Suspend new or expectant mothers from work in specific cases
- Ensure protection for young workers and not employ young workers in specific cases.

Duties of employees
- Use any equipment in accordance with any training and instructions
- Inform employer of work situations which represent serious or immediate danger

- Inform employer of shortcomings in protection arrangements.

Schedule 1: general principles of prevention

In implementing preventive and protective measures, the following hierarchy of measures must be considered:

(a) avoiding risks;
(b) evaluating the risks which cannot be avoided;
(c) combating the risks at source;
(d) adapting the work to the individual, especially as regards the design of workplaces, the choice of work equipment and the choice of working and production methods, with a view, in particular, to alleviating monotonous work and work at a predetermined work rate and to reducing their effect on health;
(e) adapting to technical progress;
(f) replacing the dangerous by the non-dangerous or the less dangerous;
(g) developing a coherent overall prevention policy which covers technology, organisation of work, working conditions, social relationships and the influence of factors relating to the working environment;
(h) giving collective protective measures priority over individual protective measures; and
(i) giving appropriate instructions to employees.

 1(d) Approved codes of practice

Management of health and safety at work

 1(e) HSE guidance notes

Five steps to risk assessment
Managing crowds safely
Managing health and safety on work experience
New and expectant mothers at work
Reducing error and influencing behaviour
Successful health and safety management
Young people at work

 3(a) Tables and figures

Key elements of successful health and safety management

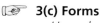 **3(c) Forms**
Hazard report

Manual Handling Operations Regulations 1992

Responsible for enforcement
    HSE and local authorities
Level of duty
    So far as is reasonably practicable
Duties of employers
- Avoid the need for employees to undertake manual handling operations which involve a risk of injury
- Where it is not reasonably practicable to avoid manual handling, make a suitable and sufficient assessment having regard to the factors and questions outlined in Schedule 1
- Take appropriate steps to reduce the risk to the lowest level
- Take appropriate steps to provide employees with general indications and where it is reasonably practicable to do so, precise information on the weight of each load and the heaviest side of any load whose centre of gravity is no longer positioned centrally.

Duties of employees
- Make full and proper use of any system of work provided.

Schedule 1 – Factors for consideration when making an assessment of manual handling operations
- the tasks
- the loads
- the working environment
- individual capability
- other factors e.g. effects of protective clothing.

 1(e) HSE guidance notes
    Manual handling
    Manual handling: solutions you can handle
 2(b) Hazard checklists
    Manual handling operations

3(a) Tables and figures
    Manual handling – lifting and lowering
    Manual Handling Operations Regulations 1992 – Flow chart

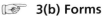 **3(b) Forms**
    Manual Handling Operations Regulations 1992 – Risk assessment

Personal Protective Equipment at Work Regulations 1992

Responsible for enforcement
    HSE and local authorities
Level of duty
    Absolute
Duties of employers
- Ensure that suitable PPE is provided except where and to the extent that the risk has been controlled by other means
- To be suitable, PPE must:
  - (a) be appropriate for the risks involved and the conditions at the place where exposure occurs
  - (b) take account of the ergonomic requirements and state of health of the wearer or user
  - (c) be capable of fitting the wearer correctly
  - (d) be effective to adequately prevent or control the risks without increasing the overall risk and
  - (e) comply with any enactment on design or manufacture listed in Schedule 1
- Where employees need to wear or use more than one item of PPE simultaneously (due to more than one risk), such equipment must be compatible and continue to be effective against the risks in question
- Before choosing any PPE that is required to be provided, ensure an assessment is made to determine whether the PPE to be provided is suitable
- The assessment shall include:
  - (a) an assessment of any risks which have not been avoided by other means

(b) the definition of the characteristics which the PPE must have in order to be effective, taking into account any risks which the PPE itself may create

(c) comparison of the PPE available with the necessary characteristics above

- Ensure that PPE provided is maintained (including replaced or cleaned as appropriate) in an efficient state, in efficient working order and in good repair
- Ensure that appropriate accommodation is provided for PPE when it is not being used
- Provide information, instruction and training as is adequate to enable the employees to know:
  (a) the risk(s) which the PPE will avoid or limit
  (b) the purpose for which and the manner in which the PPE is to be used and
  (c) action to ensure PPE remains in an efficient state, etc.
- Take all reasonable steps to ensure that any PPE is properly used by employees.

Duties of employees

- Use any PPE provided both in accordance with any training and the instructions respecting that use
- Take all reasonable steps to ensure PPE is returned to the accommodation provided after its use
- Forthwith report to his employer any loss of or obvious defect in that PPE.

☞ **1(e) HSE guidance notes**
   Personal protective equipment at work: guidance on regulations
   Respiratory protective equipment: a practical guide for users
   The selection, use and maintenance of respiratory protective equipment

☞ **2(b) Hazard checklists**
   Personal protective equipment

☞ **3(a) Tables and figures**
   Personal Protective Equipment at Work Regulations 1992 – Specimen risk survey table for the use of personal protective equipment

Pressure Systems Safety Regulations 2000

Responsible for
  HSE
Level of duty
  Absolute
Defence
  Act or default of another person
  All reasonable precautions and all due diligence
Duties of designers, manufacturers and suppliers
  • Ensure proper design and manufacture, constructed from suitable material so designed as to prevent danger, and constructed that all necessary examinations can be carried out
  • Design and construct as to ensure that access can be gained without danger (P)
  • Provide with necessary protective devices
  • Provide sufficient written information concerning design, construction, examination, operation and maintenance as may reasonably foreseeably be needed to comply with regulations
  • Manufacturers of pressure vessels to ensure information in Schedule 3 is marked on the vessel.
Duties of importers
  • Must not import a pressure vessel unless it is so marked.
Duties of persons
  • Must not remove from a pressure vessel any mark or plate containing information specified in Schedule 3
  • Must not falsify any mark on a pressure system relating to its design, construction, test or operation
  • Must not draw up a written scheme of examination unless it is suitable and:
    ○ specifies the nature and frequency of examination
    ○ specifies any measures necessary to prepare the system for safe examination and
    ○ where appropriate, provides for an examination to be carried out before the system is used for the first time.

Duties of employers of persons who undertake work

- Ensure safe installation as not to give rise to danger or otherwise impair the operation of any protective device or inspection facility
- Ensure that nothing about the way in which a pressure system is modified or repaired gives rise to danger.

Duties of users of installed systems and owners of mobile systems

- Must not operate the system or allow it to be operated unless the safe operating limits have been established
- Must have a written scheme for the periodic examination, by a competent person, of the following parts:
  ○ all protective devices
  ○ every pressure vessel and every pipeline in which a defect may give rise to danger
  ○ those parts of the pipework in which a defect may give rise to danger
- Ensure parts of system included in the scheme of examination are examined by a competent person within the intervals specified
- Before each examination, take appropriate safety measures
- Ensure system is not operated unless repairs or modifications specified in written report have been made
- Provide adequate and suitable instructions for operators and ensure operation according to the instructions
- Ensure system is properly maintained in good repair
- Keep the last report of examination by a competent person and previous reports containing appropriate information
- Give to a new user/owner copies of records

Duties of competent persons

- Undertake examinations properly and in accordance with the scheme of examination
- Following examination, make a written report, and send it to the user or owner within 28 days
- Where of the opinion that a pressure system will give rise to imminent danger unless repairs or modifications have been carried out, or suitable changes to operating conditions made, make a written report to the user/owner, sending same particulars to the enforcing authority.

1(d) Approved codes of practice

Pressure systems

 1(e) HSE guidance notes

Safety at autoclaves
Safety in pressure testing
The assessment of pressure vessels operating at low temperature

Provision and Use of Work Equipment Regulations 1998

Responsible for
    HSE and local authorities
Level of duty
    Absolute (except where stated)
Duties of employers

- Ensure that work equipment is so constructed or adapted as to be suitable for the purpose for which it is used or provided
- In selecting work equipment, have regard to the working conditions, the risks which exist in the premises or under-taking and any additional risk posed by the use of that work equipment
- Work equipment to be used only for operations for which, and under conditions for which, it is suitable (suitable means suitable in any respect which it is reasonably fore-seeable will affect the health or safety of any person)
- Ensure work equipment is maintained in an efficient state, in efficient working order and in good repair
- Where any machinery has a maintenance log, it must be kept up to date
- Ensure that, where the safety of work equipment depends upon the installation conditions, it is inspected:
  - after installation and before being put into service for the first time or
  - after assembly at a new site or in a new location

- ○ to ensure that it has been installed correctly and is safe to operate
- Where work equipment is exposed to conditions causing deterioration, which is liable to result in dangerous situations, it must be inspected at suitable intervals and at each time that exceptional circumstances liable to jeopardise its safety have occurred
- Inspection in the above cases means such visual or more rigorous inspection by a competent person and, where appropriate to carry out testing for the purpose, includes testing the nature and extent of which are appropriate for the purpose
- Ensure that the result of an inspection is recorded and kept until the next inspection is recorded
- Ensure that no work equipment leaves their undertaking or, if obtained from the undertaking of another person, is used in his undertaking unless it is accompanied by physical evidence that the last inspection has been carried out
- Where the use of work equipment is likely to involve a specific risk, ensure that the use of that equipment is restricted to persons given the task of using it and repairs, modifications, maintenance or servicing are restricted to specifically designated persons (designated persons must receive adequate training related to the operations for which they have been designated)
- Ensure that all persons who use work equipment, or who supervise or manage the use of same, have adequate health and safety information and, where appropriate, written instructions
- Information and instructions shall include information and, where appropriate, written instructions on the conditions in which and the methods by which the equipment may be used, foreseeable abnormal situations and the action to be taken, and any conclusions to be drawn from experience of using the equipment
- Information and instructions must be readily comprehensible to those concerned

- Ensure that all persons using work equipment, and those who supervise or manage the use of work equipment, have received adequate training, including training in the methods which may be adopted when using the equipment, any risks which such use may entail and the precautions to be taken
- Ensure that an item of work equipment has been designed and constructed in compliance with any essential requirements, that is, requirements relating to its design or construction in any of the instruments listed in Schedule 1 (being instruments which give effect to Community Directives concerning the safety of products)
- Ensure that measures are taken in accordance with the paragraph below which are effective:
    - to prevent access to any dangerous part of machinery or to any rotating stock-bar or
    - to stop the movement of any dangerous part of machinery or rotating stock-bar before any part of a person enters a danger zone (Danger zone means any zone in and around machinery in which a person is exposed to a risk to health and safety from contact with a dangerous part of machinery or a rotating stock-bar; stock-bar means any part of a stock-bar which projects beyond the head-stock of a lathe)
- Measures required by the above paragraph consist of
    - the provision of fixed guards enclosing every dangerous part, where and to the extent that it is practicable to do so, but where or to the extent that it is not, then
    - the provision of other guards or protection devices, where and to the extent that it is practicable to do so, but where or to the extent that it is not, then
    - the provision of jigs, holders, push-sticks and similar protection appliances used in conjunction with the machinery, where and to the extent that it is practicable to do so, but where or to the extent that it is not
    - the provision of information, instruction, training and supervision

- All guards and protection devices shall:
    - be suitable for the purpose for which they are provided
    - be of good construction, sound material and adequate strength
    - be maintained in an efficient state, in efficient working order and in good repair
    - not give rise to any increased risk to health or safety
    - not be easily bypassed or disabled
    - be situated at sufficient distance from the danger zone
    - not unduly restrict the view of the operating cycle of the machinery, where such a view is necessary and
    - be so constructed or adapted that they allow operations necessary to fit or replace parts and for maintenance work, restricting access so that it is allowed only to the areas where the work is to be carried out and, if possible, without having to dismantle the guard or protection device
- Take measures to ensure that exposure of employees to work equipment hazards is prevented or, where this is not reasonably practicable, adequately controlled. The measures required shall be other than the provision of PPE or of information, instruction and training. Specified hazards are:
    - any article or substance falling or being ejected from work equipment
    - rupture or disintegration of parts of work equipment
    - work equipment catching fire or overheating
    - the unintended or premature discharge of any article or of any gas, dust, etc. produced, used or stored in it
    - the unintended or premature explosion of the work equipment or any article or substance produced, used or stored in it
- Ensure that work equipment, parts of same and any article or substance produced, used or stored in it which is at a high or very low temperature is protected so as to prevent injury by burn, scald or sear

- Ensure that, where appropriate, work equipment is provided with one or more controls for starting (including re-starting after a stoppage) or controlling any change in speed, pressure or other operating conditions where such conditions after the change results in risks. Where such a control is required, it must not be possible to perform these operations except by a deliberate action on same
- Where appropriate, equipment must be provided with one or more readily accessible controls the operation of which will bring the work equipment to a safe condition in a safe manner
- Where appropriate, equipment must be provided with one or more readily accessible emergency stop controls unless it is not necessary by virtue of the nature of the hazards and the time taken for the work equipment to come to a complete stop as the result of the action of any stop control
- All controls must be clearly visible and identifiable, including appropriate marking where necessary. No control should be in a position where the operator could be exposed to a risk
- Ensure that control systems are safe and are chosen making due allowance for failures, faults and constraints to be expected in the planned circumstances of use (RP)
- A control system shall not be safe unless:
    - its operation does not create any increased risk
    - it ensures, so far as is reasonably practicable, that any fault in or damage to the system or loss of supply of energy cannot result in additional or increased risk and
    - it does not impede the operation of any stop control or emergency stop control
- Ensure that, where appropriate, work equipment is provided with suitable means to isolate it from all sources of energy. The means must be clearly identifiable and readily accessible
- Re-connection to an energy source must not create risks to health or safety
- Ensure stabilisation by clamping or otherwise where necessary

- Provide suitable and sufficient lighting, which takes account of the operations to be carried out
- Take appropriate measures to ensure that work equipment is so constructed or adapted that maintenance operations which involve a risk can be carried out while the equipment is shut down, or in other cases, carried out without exposing persons involved to a risk, or appropriate protective measures are taken (RP)
- Mark work equipment in a clearly visible manner
- Incorporate any warnings or warning devices which are appropriate; warnings must be unambiguous, easily perceived and easily understood
- No employee must be carried on mobile work equipment unless it is suitable for carrying persons and it incorporates features for reducing so far as is reasonably practicable risks to their safety, including risks from wheels or tracks
- Where there is a risk from mobile equipment rolling over, it must be minimised by stabilising same, a structure which ensures that the equipment does no more than fall on its side, a structure giving sufficient clearance to anyone being carried if it overturns further than that, or a device giving comparable protection
- Where there is a risk of being crushed by its rolling over, the rider shall be provided with a suitable restraining system
- In the case of a fork lift truck, this must be adapted or equipped to reduce to as low as is reasonably practicable the risk to safety from its overturning
- Where self-propelled work equipment may, while in motion, involve risk to the safety of persons:
    ○ it must incorporate facilities for preventing it being started by an unauthorised person
    ○ it must incorporate facilities for minimising the consequences of a collision where there is more than one item of rail-mounted equipment in motion at the same time
    ○ it must incorporate a device for braking and stopping

- ○ emergency braking and stopping facilities must be available in the event of failure of the main facility
- ○ where the driver's field of vision is inadequate to secure safety, adequate devices for improving his vision must be provided
- ○ if provided for use at night or in dark places, it must be equipped with lights and safe for use and
- ○ if it, or anything carried or towed by it, constitutes a fire hazard, it must carry appropriate fire-fighting equipment, unless such equipment is kept sufficiently close to it

- Where such equipment involves a risk while in motion, employers must ensure that it stops automatically once it leaves its control range and, where the risk is of crushing or impact, it incorporates features to guard against such risk, unless devices fitted are able to do so
- Where the seizure of the drive shaft between mobile work equipment and its accessories or anything towed is likely to involve a risk, employers must ensure the equipment has a means of preventing such seizure or, where seizure cannot be avoided, take every possible measure to avoid an adverse effect on the safety of an employee
- Where mobile work equipment has a shaft for the transmission of energy and the shaft could become soiled or damaged by contact with the ground while uncoupled, the equipment has a system for safeguarding the shaft.

☞ **1(d) Approved codes of practice**
Safe use of work equipment
☞ **1(e) HSE guidance notes**
Application of electro-sensitive protective equipment using light curtains and light beam devices to machinery
Drilling machines
Health and safety in engineering workshops
Power presses: maintenance and thorough examination
Safety in the use of abrasive wheels
Safety in working with lift trucks
Safeguarding of agricultural machinery

Safe use of power presses
Safe use of woodworking machinery
Safe use of work equipment: Guidance on the Regulations
Safe work with overhead travelling cranes

☞ **2(b) Hazard checklists**
Mobile mechanical handling equipment (lift trucks, etc)
Work equipment

Regulatory Reform (Fire Safety) Order 2005

Responsible for enforcement
Area fire authority
HSE (nuclear installations, ships of HM Navy, construction sites)
Fire service maintained by Secretary of State for Defence
Relevant local authority (sports grounds, regulated stands)
Fire inspectors (Crown premises, UKAEA premises)
Level of duty
Absolute (in some cases, RP)
Defence
All reasonable precautions and all due diligence
Onus of proving limits of what is reasonably practicable
In any proceedings for an offence consisting of a duty or requirement so far as is reasonably practicable, it is for the accused to prove that it was not reasonably practicable to do more than was in fact done to satisfy the duty of requirement.
Duties of responsible persons
• Take such general fire precautions as will ensure safety of any of his employees (RP). In relation to relevant persons who are not his employees, take such general fire precautions as may reasonably be required in the circumstances of the case to ensure that the premises are safe
• Make a suitable and sufficient assessment of the risks to which relevant persons are exposed for the purpose of identifying the general fire precautions he needs to take to comply with the requirements and prohibitions imposed upon him by or under this Order

- Consider implications of presence of dangerous substances in the risk assessment process
- Review risk assessment if no longer valid or there has been a significant change in the matters to which it relates
- Not employ a young person unless he has considered matters to be taken into particular account set out in Part 2 of Schedule 1
- Record the significant findings of the risk assessment and details of any group being especially at risk
- Not to commence a new work activity involving a dangerous substance unless a risk assessment has been made and measures required by the Order have been implemented
- When implementing preventive and protective measures to do so on the basis of the principles specified in Part 3 of Schedule 1
- Make and give effect to arrangements for the effective planning, organisation, control, monitoring and review of preventive and protective measures
- Record the arrangements in specified cases
- Where a dangerous substance is present, eliminate or reduce risks (RP)
- Replace a dangerous substance or use of a dangerous substance with a substance or process which eliminates or reduces risks (RP)
- Where not RP to reduce above risks, apply measures to control the risk and mitigate the detrimental effects of fire
- Arrange safe handling, storage and transport of dangerous substances and wastes
- Ensure any conditions necessary for eliminating or reducing risk are maintained
- Ensure premises are equipped with appropriate fire-fighting equipment and with fire detectors and alarms and that non-automatic fire-fighting equipment is easily accessible, simple to use and indicated by signs
- Take measures for fire-fighting in the premises, nominate competent persons to implement these measures and arrange any necessary contact with external services

- Ensure routes to emergency exits and the exits themselves are kept clear at all times
- Comply with specific requirements dealing with emergency routes, exits and doors and the illumination of emergency routes and exits in respect of premises
- Establish and, where necessary, give effect to appropriate procedures for serious and imminent danger and for danger zones, including safety drills, nomination of competent persons to implement the procedures and restriction of access to areas on the grounds of safety
- Ensure additional emergency measures are taken in respect of dangerous substances, including provision of information on emergency arrangements, suitable warnings and other communication systems, before any explosion conditions are reached, visual and audible warnings, and escape facilities
- Relevant information must be made available to emergency services and displayed at the premises
- In the event of an accident, incident or emergency related to the presence of a dangerous substance, take immediate steps to mitigate the effects of fire, restore the situation to normal, and inform relevant persons
- Ensure only those persons essential for the carrying out of repairs and other necessary work are permitted in an affected area
- Ensure that premises and any facilities, equipment and devices are subject to a suitable system of maintenance and are maintained in an efficient state, in efficient working order and in good repair
- Appoint one or more competent persons to assist him in undertaking the preventive and protective measures, ensuring adequate co-operation between competent persons
- Ensure that competent persons have sufficient time to fulfil their functions and the means at their disposal are adequate having regard to the size of the premises, the risks and the distribution of those risks

- Ensure competent persons not in his employment are informed of factors affecting the safety of any person and are provided with the same information as employees
- Provide employees with comprehensible and relevant information on the risks identified in the risk assessment, preventive and protective measures, the identities of competent persons for the purposes of evacuation of premises and the notified risks arising in shared workplaces
- Before employing a child, provide the parent with comprehensible and relevant information on the risks to that child, the preventive and protective measures and the notified risks arising in shared workplaces
- Where a dangerous substance is on the premises, provide employees with the details of any such substance and the significant findings of the risk assessment
- Provide information to employers and the self-employed from outside undertakings with respect to the risks to those employees and the preventive and protective measures taken
- Provide non-employees working in his undertaking with appropriate instructions and comprehensible and relevant information regarding any risks to those persons
- Ensure the employer of any employees from an outside undertaking working in or on the premises is provided with sufficient information with respect to evacuation procedures and the competent persons nominated to undertake evacuation procedures
- Ensure employees are provided with adequate safety training at the time of first employment, and on being exposed to new or increased risks arising from transfer or change of responsibilities, introduction of, or change in, work equipment, the introduction of new technology and the introduction of a new system of work or a change respecting an existing system of work
- In the case of shared workplaces, to co-operate with other responsible person(s), take all reasonable steps to co-ordinate the measures he takes to comply with this Order with the measures taken by other responsible persons, and take all reasonable steps to inform other responsible persons.

Duties of employees
- Every employee must:
    - take reasonable care for the safety of himself and others who may be affected by his acts or omissions while at work
    - co-operate with his employer to enable him to comply with any duty or requirement imposed by this Order
    - inform his employer or any other employee with the specific responsibility for the safety of his fellow employees of any work situation which represents a serious and immediate danger to safety, and any other matter which represents a shortcoming in the employer's protection arrangements for safety.

Powers of the Secretary of State
- The Secretary of State may by regulations make provision as to the precautions which are to be taken or observed in relation to the risk to relevant persons as regards premises to which this Order applies.

Enforcement of Order
- Every enforcing authority must enforce the provisions of this Order and any regulations made under it
- The enforcing authority must have regard to such guidance as the Secretary of State may give.

Enforcing authorities – Powers of inspectors
- An inspector may do anything necessary for the purpose of carrying out this Order and any regulations made under it into effect and, in particular, so far as may be necessary for that purpose, shall have the power to do at any reasonable time the following:
    - to enter premises and to inspect the whole or part of that premises
    - to make such inquiry as may be necessary
    - to ascertain as regards the premises, whether the provisions of this Order or regulations made under it apply or have been complied with
    - to identify the person responsible in relation to the premises
    - to require the production of any records

- ○ to require any person having responsibilities in relation to any premises to give him such reasonable facilities and assistance
- ○ to take samples of articles and substances found in any premises for the purpose of ascertaining their fire resistance or flammability
- ○ to cause any article or substance found in any premises to be dismantled or subjected to any process or test
- An inspector must, if so required, produce evidence of his authority
- Where intending to cause any article or substance to be dismantled or subjected to any process or test, at the request of a person present at the time, to cause anything which is to be done in the presence of that person
- An inspector must consult the above person(s) for the purposes of ascertaining what dangers, if any, there may be in doing anything which he proposes to do under that power
- The above powers conferred on a fire inspector, or any other person authorised by the Secretary of State, are also exercisable by an officer of the fire brigade maintained by the fire authority when authorised in writing by such an inspector.

Alterations Notices

- Where premises constitute a serious risk to relevant persons or may constitute such a risk if any change is made to them or the use to which they are put, the enforcing authority (EA) may serve on the responsible person an Alterations Notice
- Where an Alterations Notice has been served, before making any of the following changes which may result in a significant increase in risk, namely:
  - ○ a change to the premises
  - ○ a change to the services, fittings or equipment in or on the premises
  - ○ an increase in the quantities of dangerous substances which are present in or on the premises
  - ○ a change to the use of the premises

the responsible person must notify the EA of the proposed changes.

Enforcement Notices

- If the EA is of the opinion that the responsible person has failed to comply with any provision of this Order or of any regulations made under it, the enforcing authority may serve on that person an Enforcement Notice
- An Enforcement Notice may include directions as to the measures which the EA consider are necessary to remedy the above failure, including a choice between different ways of remedying the contravention
- A court may cancel or modify an Enforcement Notice
- An EA may withdraw a notice at any time before the end of the period specified, or extend or further extend the period of the notice.

Prohibition Notices

- If the EA is of the opinion that use of premises involves or will involve a risk to relevant persons so serious that use of the premises ought to be prohibited or restricted, the authority may serve on the responsible person a Prohibition Notice, such a Notice to include anything affecting the escape of relevant persons from the premises
- A Prohibition Notice must:
  - state that EA is of the opinion referred to above
  - specify the matters which give or will give rise to that risk
  - direct that the use to which the notice relates is prohibited or restricted to such extent as may be specified until the specified matters have been remedied
- A Prohibition Notice may take immediate effect or be deferred for a period specified in the notice
- Before serving a Prohibition Notice in relation to a house in multiple occupation the EA shall, where practicable, notify the local housing authority.

Appeals

- A person on whom an Alterations Notice, an Enforcement Notice, a Prohibition Notice or a notice given by the fire authority respecting fire-fighter's switches for luminous signs is served may, within 21 days, appeal to a Magistrates' Court

- On appeal, the court may either cancel or affirm the notice in its original form or with modifications
- Where an appeal is brought against an Alterations Notice or an Enforcement Notice, such appeal has the effect of suspending the operation of the notice
- Where an appeal is brought against a Prohibition Notice, such appeal does not have the effect of suspending the notice, unless the court so directs
- A person, and the EA, if aggrieved by an order made by a Magistrates Court, may appeal to the Crown Court.

Miscellaneous

- Certain luminous tube signs designed to work at a voltage normally exceeding the prescribed voltage, or other equipment so designed, must be provided with a cut-off switch so placed and coloured or marked as to be readily recognisable and accessible to fire-fighters
- The responsible person must ensure that the premises and any facilities, equipment and devices for the use by or protection of fire-fighters are subject to a suitable system of maintenance and are maintained in an efficient state, in efficient working order and in good repair
- Nothing in this Order is to be construed as conferring a right of action in any civil proceedings (other than proceedings for the recovery of a fine)
- Breach of a duty imposed on an employer by or under this Order, so far as it causes damage to an employee, confers a right of action on that employee in any civil proceedings
- No employer must levy or permit to be levied on any employee of his any charge in respect of anything done or provided in pursuance of any requirement of this Order or regulations made under this Order.
- In the case of licensed premises:
  - the licensing authority must consult the EA before issuing the licence

○ the EA must notify the licensing authority of any action that the EA takes
- Where it is proposed to erect a building, or make any extension of or structural alteration to a building to which the Order applies, the local authority must consult the EA before passing those plans.

Service of notices
- Similar provisions as those for the HSWA apply with respect to the service of notices.

Schedule 1
Part I – Matters to be considered in risk assessment in respect of dangerous substances
- The matters are:
    ○ the hazardous properties of the substance
    ○ information on safety provided by the supplier, including information contained in any relevant safety data sheet
    ○ the circumstances of the work including:
        (i) the special, technical and organisational measures and the substances used and their possible interactions
        (ii) the amount of the substance involved
        (iii) where the work will involve more than one dangerous substance, the risk presented by such substances in combination
        (iv) the arrangements for the safe handling, storage and transport of dangerous substances and of waste containing dangerous substances
        (v) activities, such as maintenance, where there is the potential for a high level of risk
        (vi) the effect of measures which have been or will be taken pursuant to this Order
        (vii) the likelihood that an explosive atmosphere will occur and its persistence
        (viii) the likelihood that ignition sources, including electrostatic discharges, will be present and become active and effective

(ix) the scale of the anticipated effects

(x) any places which are, or can be connected via openings to, places in which explosive atmospheres may occur

(xi) such additional safety information as the responsible person may need in order to complete the assessment.

Part 2 – Matters to be taken into particular account in risk assessment in respect of young persons
- The matters are:
    ○ the inexperience, lack of awareness of risks and immaturity of young persons
    ○ the fitting-out and layout of the premises
    ○ the nature, degree and duration of exposure to physical and chemical agents
    ○ the form, range and use of work equipment and the way in which it is handled
    ○ the organisation of processes and activities
    ○ the extent of the safety training provided or to be provided to young persons
    ○ risks from agents, processes and work listed in the Annex to Council Directive 94/33/EC on the protection of young people at work.

Part 3 – Principles of prevention
- These principles are:
    ○ avoiding risks
    ○ evaluating the risks which cannot be avoided
    ○ combating the risks at source
    ○ adapting to technical progress
    ○ replacing the dangerous by the non-dangerous or less-dangerous
    ○ developing a coherent overall prevention policy which covers technology, organisation of work and the influence of factors relating to the working environment
    ○ giving collective protective measures priority over individual protective measures
    ○ giving appropriate instructions to employees.

Part 4 – Measures to be taken in respect of dangerous substances

- In applying measures to control risks the responsible person must, in order of priority:
    - reduce the quantity of dangerous substances to a minimum
    - avoid or minimise the release of a dangerous substance
    - control the release of a dangerous substance at source
    - prevent the formation of an explosive atmosphere, including the application of appropriate ventilation
    - ensure that any release of a dangerous substance which may give rise to risk is suitably collected, safely contained, removed to a safe place or otherwise rendered safe, as appropriate
    - avoid:
        - (i) ignition sources including electrostatic discharges
        - (ii) such other adverse conditions as could result in harmful physical effects from a dangerous substance
    - segregate incompatible dangerous materials
- The responsible person must ensure that mitigation measures applied in accordance with article 12(3)(b) include:
    - reducing to a minimum the number of persons exposed
    - measures to avoid the propagation of fires or explosions
    - providing explosion pressure relief arrangements
    - providing explosion suppression equipment
    - providing plant which is constructed so as to withstand the pressure likely to be produced by an explosion
    - providing personal protective equipment
- The responsible person must:
    - ensure that the premises are designed, constructed and maintained so as to reduce risk
    - ensure that suitable special, technical and organisational measures are designed, constructed, assembled, installed, provided and used so as to reduce risk

- ○ ensure that special, technical and organisational measures are maintained in an efficient state, in efficient working order and in good repair
- ○ ensure that equipment and protective systems meet the following requirements:
    - (i) where power failure can give rise to the spread of additional risk, equipment and protective systems must be able to be maintained in a safe state of operation independently of the rest of the plant in the event of power failure
    - (ii) means for manual override must be possible, operated by employees competent to do so, for shutting down equipment and protective systems incorporated within automatic processes which deviate from the intended operating conditions, provided that the provision or use of such means does not compromise safety
    - (iii) on operation of emergency shutdown, accumulated energy must be dissipated as quickly and as safely as possible or isolated so that it no longer constitutes a hazard and
    - (iv) necessary measures must be taken to prevent confusion between connecting devices
- ○ where the work is carried out in hazardous places or involves hazardous activities, ensure that appropriate systems of work are applied including:
    - (i) the issuing of written instructions for carrying out the work and
    - (ii) a system of permits to work, with such permits being issued by a person with responsibility for this function prior to commencement of the work concerned.

☞ 1(e) HSE guidance notes

The safe use of compressed gases in welding, flame cutting and allied processes
The storage of flammable liquids in containers
The storage of flammable liquids in tanks

2(b) Hazard checklists
Fire safety
 3(a) Tables and figures
Fire instruction notice

Reporting of Injuries, Diseases and Dangerous Occurrences Regulations 1995

Responsible for enforcement
   HSE and local authorities
Level of duty
   Absolute
Duties of responsible persons
   • Notify the relevant enforcing authority by quickest practicable means and make a report on the approved form in the case of:
      ○ the death of any person as a result of an accident arising out of or in connection with work any person at work suffering a specified major injury
      ○ any person who is not at work suffering an injury as a result of an accident arising out of or in connection with work and where that person is taken to hospital for treatment
      ○ any person who is not at work suffering a major injury as a result of an accident arising out of or in connection with work at a hospital
      ○ where there is a dangerous occurrence
   • As soon as practicable, and within 10 days, report any situation where a person at work is incapacitated for more than 3 consecutive days (excluding the day of the accident but including any days which would not have been working days) because of an injury resulting from an accident arising out of or in connection with work
   • Where an employee has suffered a reportable injury which is a cause of his death within one year of the date

of the accident, inform the relevant enforcing authority as soon as it comes to his knowledge

- Where:
  - ○ a person at work
  - ○ a person at an offshore workplace

  suffers a scheduled occupational disease, send a report to the relevant enforcing authority
- Keep records of all reportable injuries, diseases and dangerous occurrences.

Duties of conveyors of flammable gas through a fixed pipe distribution system and of fillers, importers and suppliers of refillable containers containing LPG

- Report death or major injury which has arisen out of or in connection with the gas distributed, filled, imported or supplied within 14 days to the HSE.

Duties of CORGI-registered gas installers

- Where a gas fitting or flue or ventilation used in connection with that fitting is or has been likely to cause death or any major injury by reason of:
  - ○ accidental leakage of gas
  - ○ inadequate combustion of gas
  - ○ inadequate removal of the products of combustion of gas

  within 14 days, send a report to the HSE
- Report forms
  Form 2508   Report of an injury or dangerous occurrence
  Form 2508A   Report of a case of disease
  Form 2508G   Report of a gas incident.

Notifiable and reportable major injuries

- Any fracture other than to the fingers, thumbs or toes
- Any amputation
- Dislocation of the hip, knee or spine
- Loss of sight (whether temporary or permanent)
- A chemical or hot metal burn to the eye
- Any injury resulting from electric shock or electrical burn (including any electrical burn caused by arcing or arcing products) leading to unconsciousness or requiring resuscitation or admittance to hospital for more than 24 hours

- Any other injury:
  - (a) leading to hypothermia, heat-induced illness or to unconsciousness
  - (b) requiring resuscitation or
  - (c) requiring admittance to hospital for more than 24 hours
- Loss of consciousness caused by asphyxia or by exposure to a harmful substance or biological agent
- Either of the following conditions which result from the absorption of any substance by inhalation, ingestion or through the skin:
  - (a) acute illness requiring medical treatment, or
  - (b) loss of consciousness
- Acute illness which requires medical treatment where there is reason to believe that this resulted from exposure to a biological agent or its toxins or infected material.

Dangerous occurrences
- Classified under
  1. General
  2. Dangerous occurrences which are reportable in relation to mines
  3. Dangerous occurrences which are reportable in relation to quarries
  4. Dangerous occurrences which are reportable in respect of relevant transport systems
  5. Dangerous occurrences which are reportable in respect of an offshore workplace.

Reportable diseases
- Classified under
  1. Conditions due to physical agents and the physical demands of work
  2. Infections due to biological agents
  3. Conditions due to substances.

☞ **1(e) HSE guidance notes**
A guide to the Reporting of Injuries, Diseases and Dangerous Occurrences Regulations 1995

Investigating accidents and incidents
The cost of accidents at work

☞ **3(a) Tables and figures**

Reporting of Injuries, Diseases and Dangerous Occurrences Regulations 1995 – Reporting requirements

☞ **3(b) Forms**

Reporting of Injuries, Diseases and Dangerous Occurrences Regulations 1995 – Report of an injury or dangerous occurrence (Form 2508)

Reporting of Injuries, Diseases and Dangerous Occurrences Regulations 1995 – Report of a case of disease (Form 2508A)

Safety Representatives and Safety Committees Regulations 1977

Responsible for enforcement
HSE and local authorities
Level of duty
Absolute
Defence
None
Functions of a trade union
- To appoint safety representatives from amongst the employees of a recognised trade union
- To notify the employer in writing the names of the persons appointed as safety representatives and the group or groups of employees they represent
- To terminate the appointments of safety representatives.

Functions of safety representatives
- To investigate potential hazards and dangerous occurrences and to examine the cause of accidents at the workplace
- To investigate complaints by any employee he represents relating to that employee's health, safety or welfare at work
- To make representation to the employer on matters arising from the above functions

- To make representations to the employer on general matters affecting the health, safety or welfare at work of the employees at the workplace
- To carry out inspections of the workplace
  (a) on a frequent basis
  (b) where there has been a substantial change in the conditions of work
  (c) where new information has been published by the HSC or HSE
  (d) where there has been a notifiable accident or dangerous occurrence in a workplace
- To inspect and take copies of documents relevant to the workplace or to the employees the safety representative represents which the employer is required to keep
- To receive information which, within the employer's knowledge, is necessary to enable them to fulfil their functions
- To represent the employees he was appointed to represent in consultations at the workplace with inspectors of the HSE and of any other enforcing authority
- To receive information from inspectors
- To request the establishment of a safety committee
- To attend meetings of safety committees
- To give the employer reasonable notice in writing prior to workplace inspections.

Duties of employers

- To permit safety representatives to take such time off with pay during working hours as necessary for:
  (a) performing the above functions; and
  (b) undergoing such training in aspects of those functions as may be reasonable in all the circumstances
- To provide facilities and assistance for the purpose of carrying out inspections
- When requested by safety representatives, to establish a safety committee not later than 3 months following the request.

 1(d) Approved codes of practice
Safety representatives and safety committees

1(e) HSE guidance notes
Investigating accidents and incidents
Safety representatives and safety committees

Safety Signs Regulations 1980

Responsible for enforcement
HSE, local authorities
Level of duty
Absolute
Duties of employers
- Any sign displayed in the workplace must comply with the specification of signs contained in BS 5378: Part 1: 1980 Safety Signs and Colours: Specifications for Colour and Design Classification of signs:
  - Prohibition e.g. No smoking
  - Warning e.g. Risk of slipping
  - Mandatory e.g. Eye protection must be worn
  - Safe condition e.g. Fire exit

 3(a) Tables and figures
Safety signs

Simple Pressure Vessels (Safety) Regulations 1991

Responsible for enforcement
HSE
Level of duty
Absolute
Duties of manufacturers
- In the case of vessels with a stored energy over 50 bar litres:
  - meet the essential safety requirements
  - have safety clearance

- bear the EC mark and other specified inscriptions
- be accompanied by manufacturer's instructions
- be safe (as defined)
- In the case of vessels with a stored energy up to 50 bar litres
  - must be manufactured in accordance with engineering practice recognised as sound in the Community country
  - bear specific inscriptions (but not the EC mark)
  - be safe
- Where he has obtained an EC certificate of conformity, may apply the CE mark to any vessels covered by the certificate where he executes an EC declaration of conformity
- Ensure the EC mark consists of the appropriate symbol, the last two digits of the year in which the mark is applied and, where appropriate, the distinguishing number assigned by the EC to the approved body responsible for EC verification or EC surveillance
- Must apply specified inscriptions to Category A and B vessels
- Duties of approved bodies (i.e approved by the Secretary for Trade and Industry)
- Where it has issued an EC verification certificate, to ensure the application of the EC mark to every vessel covered by the certificate
- Undertake EC surveillance where a certificate has been issued.

 1(e) HSE guidance notes
The assessment of pressure vessels working at low temperature

Work at Height Regulations 2005

Responsible for enforcement
    HSE and local authorities
Level of duty
    Absolute and SFARP

Defence
    None
Duties of employers

- Ensure work at height is:
    - properly planned
    - appropriately supervised and
    - carried out in a manner which is safe(SFARP)

    and that its planning includes the selection of work equipment in accordance with requirements below and for emergencies and rescue

- Ensure work at height is carried out only when the weather conditions do not jeopardise the health and safety of persons involved
- Ensure that no person engages in any activity, including organisation, planning and supervision, in relation to work at height of work equipment for use in such work, unless he is competent to do so or, if being trained, is supervised by a competent person
- Take account of risk assessment under the MHSWR
- Ensure that work is not carried out at height where it is not RP to carry out the work safely otherwise than at height
- Take suitable and sufficient measures to prevent any person falling a distance liable to cause personal injury (SFARP)
- Above measures shall include:
    - ensuring that work is carried out:
        - from an existing place of work or
        - (in the case of obtaining access or egress) using an existing means which complies with Schedule 1, where it is RP to carry it out safely and under appropriate ergonomic conditions and
        - where not RP for the work to be carried out as above, his providing sufficient work equipment for preventing a fall occurring (SFARP)
- Where the above measures do not eliminate the risk of a fall, every employer shall:
    - provide sufficient work equipment (SFARP) to minimise
        - the distance and consequences or

- – where it is not RP to minimise the distance, the consequences of a fall and
- – provide such additional training and instruction or take other additional suitable and sufficient measures to prevent any person falling a distance liable to cause injury (RP)
- In selecting work equipment for use in work at a height:
  - give collective protection measures priority over personal protection measures; and take account of:
    - – the working conditions and the risks
    - – in the case of work equipment for access or egress, the distance to be negotiated
    - – the distance and consequences of a potential fall
    - – the duration and frequency of use
    - – the need for easy and timely evacuation and rescue in an emergency
    - – any additional risk posed by the use, installation or removal of that work equipment or by
    - – evacuation and rescue from it and
    - – the other provisions of these regulations
- Select work equipment for work at height which:
  - has characteristics including dimensions which:
    - – are appropriate to the nature of the work to be performed and the foreseeable loadings and
    - – allow passage without risk
  - is in other respects the most suitable work equipment, having regard to the purposes specified above with respect to the avoidance of risks
- Ensure that, in the case of:
  - guard rail, toe board, barrier or similar collective means of protection, Schedule 2 is complied with
  - a working platform, Part 1 of Schedule 3 is complied with and
  - where scaffolding is provided, Part 2 of Schedule 3 is also complied with
  - a net, airbag or other collective safeguard for arresting falls which is not part of a personal fall protection system, Schedule 4 is complied with

- - a personal fall protection system, Part 1 of Schedule 5 is complied with
  - in the case of a work positioning system, Part 2 of Schedule 5 is complied with
  - in the case of rope access and positioning techniques, Part 3 of Schedule 5 is complied with
  - in the case of a fall arrest system, Part 4 of Schedule 5 is complied with
  - in the case of a work restraint system, Part 5 of Schedule 5 is complied with and
  - a ladder, Schedule 6 is complied with
- Ensure that no person at work passes across or near, or works on, from or near, a fragile surface where it is RP to carry out work safely and under appropriate ergonomic conditions without his doing so
- Where not RP to carry out work as above:
  - ensure that suitable and sufficient platforms, coverings, guard rails or similar means of support or protection are provided and used (RP) so that any foreseeable loading is supported by such supports or borne by such protection
  - where the risk of a person falling remains despite the above measures, take suitable and sufficient measures to minimise the distances and consequences of his fall
- Where any person may pass across or near, or work on, from or near, a fragile surface, shall ensure that:
  - prominent warning notices are affixed to the place where the fragile surface is situated (RP) or
  - where not RP, such persons are made aware of it by other means
- Take suitable and sufficient steps to prevent the fall of any material (RP)
- Where not RP to comply with the above, take suitable and sufficient steps to prevent any person being struck by any falling material or object
- Ensure no material or object is thrown or tipped from height in circumstances where it is liable to cause injury to any person

- Ensure materials and objects are stored in such a way as to prevent risk arising from the collapse, overturning or unintended movement of same
- Ensure that:
  - where a workplace contains a danger area where there is a risk of a person:
    - falling a distance or
    - being struck by a falling object

  which is liable to cause injury, the workplace is equipped with devices preventing unauthorised persons from entering such areas (RP) and such area is clearly indicated.
- In the case of:
  - guard rails, toe boards, barriers and similar collective means of protection
  - working platforms
  - scaffolding
  - collective safeguards for arresting falls
  - personal fall protection systems; and
  - ladders (see Schedules 2 to 6)

  ensure that where safety of work equipment depends on how it is installed or assembled, it is not used after installation or assembly unless it has been inspected in that position
- Ensure work equipment exposed to conditions causing deterioration which is liable to result in dangerous situations is inspected
  - at suitable intervals and
  - each time that exceptional circumstances which are liable to jeopardise the safety of the equipment have occurred
  - to ensure that health and safety conditions are maintained and that any deterioration can be detected and remedied in good time
- Ensure that a working platform:
  - used for construction work and
  - from which a person could fall 2 metres or more

  is not used in any position unless it has been inspected in that position or, in the case of a mobile working

platform, inspected on the site, within the previous 7 days
- Ensure that no work equipment, other than lifting equipment to which LOLER applies:
    - leaves his undertaking or
    - if obtained from the undertaking of another person, is used in his undertaking
    - unless it is accompanied by physical evidence that the last inspection required to be carried out has been carried out
- Ensure that an inspection is recorded and kept until the next inspection is recorded
- Keep the above report or a copy of same
    - at the site where the inspection was carried out until the construction work is completed and
    - thereafter at an office of his for 3 months
- Ensure that the surface and every parapet, permanent rail or other such fall protection measure of every place of work at height are checked on each occasion before the place is used (RP).

Duties of persons carrying out inspections
- Shall:
    - before the end of the working period within which the inspection is completed, prepare a report containing the particulars set out in Schedule 7 and
    - within 24 hours of completing the inspection, provide the report or a copy thereof to the person on whose behalf the inspection was carried out.

Duties of persons at work
- Where working under the control of another person, report to that person any activity or defect relating to work at height which he knows is liable to endanger the safety of himself or another person
- Use any work equipment or safety device provided for him in accordance with:
    - any training in the use of the work equipment or device concerned which has been received by him and

○ the instructions respecting that use which have been provided to him by that employer or person in compliance with the relevant statutory provisions.

Schedules
- Requirements for existing places of work and means of access or egress at height
- Requirements for guard rails, toe boards, barriers and similar collective means of protection
- Requirements for working platforms
  1. requirements for all working platforms
  2. additional requirements for scaffolding
- Requirements for collective safeguards for arresting falls
- Requirements for personal fall protection systems
  1. requirements for all personal fall protection systems
  2. additional requirements for work positioning systems
  3. additional requirements for rope access and positioning techniques
  4. additional requirements for fall arrest systems
- Additional requirements for work restraint systems
- Requirements for ladders
- Particulars to be incorporated in a report of inspection.

Workplace (Health, Safety and Welfare) Regulations 1992

Responsible for enforcement
HSE and local authorities
Level of duty
Absolute
Defence
None
Duties of employers
- Workplace and the equipment, devices and systems to which this regulation applies shall be maintained (which includes them being cleaned as appropriate) in an efficient state, in efficient working order and in good repair

- The equipment, devices and systems to which this regulation applies are:
  - equipment and devices that if a fault occurred in them, would then be likely to fail to comply with any of these regulations and
  - mechanical ventilation systems
- Effective and suitable provision shall be made to ensure that every enclosed workplace is ventilated by a sufficient quantity of fresh air
- During working hours, the temperature in all workplaces inside buildings shall be reasonable. (The ACOP to the regulations specifies a minimum temperature of 16°C, except where work involves severe physical effort, in which case it should be at least 13°C)
- Workplaces must have suitable and sufficient lighting, together with emergency lighting where employees may be exposed to danger in the event of the lighting system failing
- Workplaces, and the furniture, furnishings and fittings, must be kept sufficiently clean
- Floor, wall and ceiling surfaces shall be capable of being kept sufficiently clean
- Workrooms shall have sufficient floor area, height and unoccupied floor space for the purposes of health, safety and welfare. Every employee shall have a minimum space of 11 m³, and no space more than 4.2 m from the floor shall be taken into account
- Workstations shall be suitable for the persons undertaking the work and for the type of work undertaken. A suitable seat shall be provided for persons whose work, or a substantial part of it, can be done sitting
- Floors and traffic routes shall be of such construction as to be suitable for their purpose
- So far as is reasonably practicable, suitable and effective measures shall be taken to prevent any person falling a distance and being struck by a falling object, in both cases, likely to cause injury
- Windows or other transparent or translucent surfaces in walls or partitions, and transparent or translucent surfaces

in doors or gates, must be of safety material and be appropriately marked to make them apparent

- Windows, skylights and ventilators must be capable of being opened, closed or adjusted in such a manner as to prevent a person performing these operations to be exposed to risk
- Windows and skylights must be so designed or con-structed as to enable safe cleaning of same
- Workplaces must be organised in such a way as to ensure safe circulation by pedestrians and vehicles
- Doors and gates must be suitably constructed (including being fitted with any necessary safety devices)
- Escalators and moving walkways must function safely, be equipped with necessary safety devices and fitted with one or more emergency stop controls
- Suitable and sufficient sanitary conveniences must be provided at readily accessible places
- At least 1 WC must be provided for every 25 males and females (or proportion of 25)
- Suitable and sufficient washing facilities, including showers if required by the nature of the work or for health reasons, shall be provided at readily accessible places
- An adequate supply of wholesome drinking water must be provided
- Suitable and sufficient accommodation for clothing not worn during working hours, or for special clothing which is not taken home, must be provided
- Suitable and sufficient facilities shall be provided to enable employees to change clothing in cases where employees have to change clothing for the purposes of work. Separate facilities, or the separate use of facilities, shall be provided for men and women for reasons of propriety
- Suitable and sufficient rest facilities, including those for pregnant women or nursing mothers, shall be provided, together with facilities to eat meals where meals are regu-larly eaten in the workplace.

 1(d) Approved codes of practice
Workplace health, safety and welfare

☞ **1(d) HSE guidance notes**

A pain in your workplace: ergonomic problems and solutions
General ventilation in the workplace
Lighting at work
Seating at work
Slips and trips
Thermal comfort in the workplace
Workplace transport safety

☞ **2(b) Hazard checklists**

Floors and traffic routes
Maintenance work
Offices and commercial premises

☞ **3(a) Tables and figures**

Air changes per hour (comfort ventilation)
Average illuminances and minimum measured illuminances
Maximum ratios of illuminance
Optimum working temperatures
Water closets and urinals for men
Water closets and wash station provision

1(d)
Approved codes of practice

Listed below are the principal approved codes of practice issued by the Health and Safety Commission. They are published by HMSO and are available through HSE books and booksellers.

Asbestos

- Control of asbestos at work
- The management of asbestos in non-domestic premises
- Work with asbestos insulation, asbestos coating and asbestos insulation board
- Work with asbestos that does not normally require a licence

Confined spaces

- Safe work in confined spaces

Construction

- Managing health and safety in construction

Dangerous substances and explosive atmospheres

- Unloading petrol from road tankers
- Design of plant, equipment and workplaces
- Storage of dangerous substances

- Control and mitigation measures
- Safe maintenance, repair and cleaning procedures
- Dangerous substances and explosive atmospheres

Diving operations

- Commercial diving projects offshore
- Commercial diving projects inland/offshore
- Media diving projects
- Recreational diving projects
- Scientific and archaeological diving projects

Docks

- Safety in docks

First aid

- First aid at mines
- First aid at work

Gas

- Safety in the installation and use of gas systems and appliances
- Standards of training in safe gas installation
- Design, construction and installation of gas service pipes

Ionising radiation

- Work with ionising radiation

Lead

- Control of lead at work

Legionnaires' disease

- The control of Legionella bacteria in water systems

Lifting equipment

- Safe use of lifting equipment

Lift trucks

- Rider-operated lift trucks

Management of health and safety

- Management of health and safety at work

Mines

- Safety of exit from mine underground workings
- Shafts and windings in mines
- First aid at mines
- The management and administration of safety and health at mines
- Explosives at coal and other safety-lamp mines
- The prevention of inrushes in mines
- Escape and rescue from mines

- The control of ground movement in mines
- The use of electricity in mines

Offshore installations

- Prevention of fire and explosion and emergency response on offshore installations
- Health care and first aid on offshore installations and pipeline works

Pesticides

- Safe use of pesticides for non-agricultural purposes

Petroleum spirit

- Plastic containers with nominated capacities up to five litres for petroleum spirit: Requirements for testing and marking or labelling

Pottery

- Control of substances hazardous to health in the production of pottery

Power presses

- Safe use of power presses

Pressure systems

- Safety of pressure systems

Quarries

- Health and safety at quarries
- The use of electricity at quarries

Railways

- Railway safety critical work

Safety representatives and safety committees

- Safety representatives and safety committees

Substances hazardous to health

- Control of substances hazardous to health
- Control of substances hazardous to health in fumigation operations

Work equipment

- Safe use of work equipment
- Safe use of power presses
- Safe use of woodworking machinery

Workplaces

- Workplace health, safety and welfare

Zoos

- Safety, health and welfare standards for employers and persons at work

1(e)
HSE guidance notes

The range of guidance notes issued by the HSE is extensive. Guidance notes are available through HSE Books and published in the following series:

- general (G)
- chemical safety (CS)
- plant and machinery (PM)
- medical (M)
- environmental hygiene (EH)
- legal (L).

Those guidance notes which are of both more general and specific application to workplaces are listed below.

- Application of electro-sensitive protective equipment using light curtains and light beam devices to machinery [HS(G)180]
- Approved classification and labelling guide [L131]
- Asbestos essentials task manual [HS(G)210]
- Assessing and managing risks at work from skin exposure to chemical agents [HS(G)205]
- The assessment of pressure vessels operating at low temperature [HS(G)93]
- Avoiding danger from underground services [HS(G)47]
- Backs for the future: safe manual handling in construction [HS(G)149]
- Biological monitoring in the workplace [HS(G)167]
- Bulk storage of acids [HS(G)235]
- CHIP for everyone [HS(G)228]
- Choice of skin care products for the workplace [HS(G)207]
- A comprehensive guide to managing asbestos in premises [HS(G)227]
- Compressed air safety [HS(G)39]
- Control of diesel engine exhaust emissions in the workplace [HS(G)187]

- The control of Legionella in water systems [L8]
- Control of substances hazardous to health in fumigation operations
- COSHH essentials: easy steps to control chemicals [HS(G)193]
- Cost and effectiveness of chemical protective gloves for the workplace [HS(G)206]
- Dangerous goods in cargo transport units [HS(G)78]
- Display screen equipment work: guidance on Regulations [L26]
- Drilling machines [PM83]
- Dust: general principles of prevention [EH44]
- Effective health and safety training [HS(G)222]
- Electrical safety in arc welding [HS(G)118]
- Electrical safety on construction sites [HS(G)141]
- Electricity at work: safe working practices [HS(G)85]
- Fire safety in construction: guidance for clients, designers and those managing and carrying out construction work involving significant risks [HS(G)168]
- Five steps to risk assessment [HS(G)183]
- General ventilation in the workplace [HS(G)202]
- A guide to the Construction (Head Protection) Regulations 1989 [L102]
- A guide to the Gas Safety (Management) Regulations [L80]
- A guide to the Health and Safety (Consultation with Employees) Regulations 1996 [L95]
- A guide to the Reporting of Injuries, Diseases and Dangerous Occurrences Regulations 1995 [HSIS1]
- A guide to the Work in Compressed Air Regulations 1996 [L96]
- Hand-arm vibration [HS(G)88]
- Health and safety in arc welding [HS(G)204]
- Health and safety in construction [HS(G)150]
- Health and safety in engineering workshops [HS(G)129]
- Health and safety in excavations [HS(G)185]
- Health and safety in roof work [HS(G)33]
- Health risk management: a guide to working with solvents [HS(G)188]

- Health risk management: a practical guide for managers in small and medium sized enterprises [HS(G)137W]
- Health surveillance at work [HS(G)61]
- How to deal with sick building syndrome [HS(G)132]
- Introduction to asbestos essentials [HS(G)213]
- Introduction to local exhaust ventilation [HS(G)37]
- Keeping electrical switch gear safe [HS(G)230]
- The law on VDUs [HS(G)90]
- Legionnaires' disease [L8]
- Lighting at work [HS(G)38]
- Maintenance, examination and testing of local exhaust ventilation [HS(G)54]
- Maintaining portable and transportable electrical equipment [HS(G)107]
- Managing contractors [HS(G)159]
- Managing crowds safely [HS(G)154]
- Managing health and safety in construction [HS(G)224]
- Managing health and safety in dock work [HS(G)177]
- Managing health and safety in swimming pools [HS(G)179]
- Managing health and safety on work experience [HS(G)199]
- Manual handling [HS(G)115]
- Manual handling: solutions you can handle [HS(G)115]
- Memorandum of Guidance on the Electricity at Work Regulations 1989 [HS(R)25]
- Monitoring strategies for toxic substances [HS(G)173]
- New and expectant mothers at work: a guide for employers [HS(G)122]
- A pain in your workplace: ergonomic problems and solutions [HS(G)121]
- Personal protective equipment at work: guidance on Regulations [L25]
- Power presses: maintenance and thorough examination [HS(G)236]
- Preventing asthma at work [L55]
- Preventing violence to retail staff [HS(G)133]
- Prevention of violence to staff in banks or building societies [HS(G)100]
- Protecting the public: Your next move [HS(G)151]

- Reducing error and influencing behaviour [HS(G)48]
- The safe use and handling of flammable liquids [HS(G)140]
- The safe use of compressed gases in welding, flame cutting and allied processes [HS(G)139]
- Safe use of lifting equipment [L113]
- Safe use of power presses [HS(G)236]
- The safe use of vehicles on construction sites [HS(G)144]
- Safe use of woodworking machinery [L114]
- Safe work in confined spaces [L101]
- Safe work with overhead travelling cranes [PM55]
- Safeguarding agricultural machinery [HS(G)89]
- Safety at autoclaves [PM73]
- Safety in pressure testing [GS4]
- Safety in the installation and use of gas systems and appliances [L56]
- Safety in the use of abrasive wheels [HS(G)17]
- Safety in the use of pallets [PM15]
- Safety in working with lift trucks [HS(G)6]
- Safety signs and signals: Health and Safety (Safety Signs and Signals) Regulations 1996: guidance on Regulations [L64]
- Seating at work [HS(G)57]
- The selection, use and maintenance of respiratory protective equipment [HS(G)53]
- Seven steps to successful substitution of hazardous substances [HS(G)110W]
- Slips and trips [HS(G)155]
- Sound solutions: Techniques to reduce noise at work [HS(G)138]
- A step-by-step guide to COSHH assessment [HS(G)97]
- The storage of flammable liquids in containers [HS(G)51]
- The storage of flammable liquids in tanks [HS(G)176]
- Successful health and safety management [HS(G)65]
- Tackling work-related stress [HS(G)218]
- Thermal comfort in the workplace [HS(G)194]
- The training of first aid at work [HS(G)212]
- Upper limb disorders in the workplace [HS(G)60]
- Vibration solutions [HS(G)170]

- Workplace exposure limits [EH40]
- Workplace transport safety [HS(G)136]
- Work-related upper limb disorders [HS(G)60]
- Work-related violence [HS(G)229]
- Work with asbestos cement [HS(G)189/2]
- Work with display screen equipment [L26]
- Young people at work [HS(G)165]

PART 2
Health and Safety Management

2(a)
Health and safety management in practice

The duty on employers to manage health and safety is clearly specified in the Management of Health and Safety at Work Regulations. This part of the *Health and Safety Pocket Book* outlines a number of practical aspects of health and safety management.

Accident costs

All accidents represent both direct and indirect costs to employers and many organisations endeavour to calculate both the direct and indirect costs with a view to identifying future management strategies.

These costs may be summarised using the accident cost assessment form shown on next page.

Accident cost assessment form

£ p

Direct costs
% of occupier's liability premium
% of increased premiums payable

Claims
Fines and damages awarded in courts
Court and legal representation costs

Indirect costs
Treatment
    First aid
    Transport
    Hospital
    Other costs

Lost time
    Injured person
    Management
    Supervisor(s)
    First Aiders
    Other persons

Production
    Lost production
    Overtime payments
    Damage to plant, equipment, structures, vehicles, etc.
    Training and supervision of replacement labour

Investigation
    Management
    Safety adviser
    Others e.g. safety representatives
    Liaison with enforcement authority officers

Other costs
    Ex-gratia payment to injured person
    Replacement of personal items of:
        (i) injured person
        (ii) other persons
    Other miscellaneous costs

TOTAL COSTS

Accident investigation procedure

Anyone investigating an accident, particularly a fatal or major injury accident, or a scheduled dangerous occurrence, needs to follow a specific procedure.

1. Establish the facts surrounding the accident as quickly and completely as possible with respect to:
    (a) the work environment in which the accident took place, e.g. location, lighting;
    (b) the plant, machinery, equipment, and hazardous substance involved;
    (c) the system of work or working procedure involved; and
    (d) the sequence of events leading to the accident.
2. Produce sketches and diagrams of the accident scene.
3. Take photographs of the accident scene before anything is moved.
4. Identify all witnesses and make a list of witnesses.
5. Interview all witnesses in the presence of a third party and take full statements. (Witnesses should be cautioned prior to making a statement.) Do not prompt or lead the witnesses. Witnesses should agree any written statements produced, and sign and date these statements.
6. Evaluate the facts and individual witnesses' versions of the events leading to the accident with respect to accuracy, reliability and relevance.
7. Endeavour to arrive at conclusions as to both the indirect and direct causes of the accident on the basis of the relevant facts.
8. Examine closely any contradictory evidence. Never dismiss a fact that does not fit in with the rest of the facts. If necessary, find out more.
9. Examine fully the system of work in operation, in terms of the persons involved with respect to age, training, experience, level of supervision and the nature of the work, e.g. routine, sporadic or incidental.

10. In certain cases it may be necessary for plant and equipment to be examined by an expert, such as a consultant engineer.
11. Produce a written report indicating the stages prior to the accident and emphasising the causes of same. Measures to prevent a recurrence should also be incorporated in such a report. The report should be presented to the employer or his representative.
12. In complex and serious cases, it may be appropriate to establish a small investigating committee comprising the responsible manager, supervisors, safety representatives and technical specialists.
13. It should be appreciated that the thorough investigation of accidents is essential particularly where there may be the possibility of criminal proceedings by the enforcement authority and/or civil proceedings by the injured party or his representatives.

Benchmarking

A benchmark is a reference point which is commonly used in surveying practice. More recently, the term has been used to imply some form of standard against which an organisation can measure performance and, as such, is an important business improvement tool in areas such as quality management. Health and safety benchmarking follows the same principles whereby an organisation's health and safety performance can be compared with a similar organisation or 'benchmarking partner'.

The HSE publication *Health and safety benchmarking – Improving together* (IND G301/1999) defines health and safety benchmarking as 'a planned process by which an organisation compares its health and safety processes and performance with others to learn how to:
   1. reduce accidents and ill-health;

2. improve compliance with health and safety law; and/or
3. cut compliance costs.'

The benchmarking process

Health and safety benchmarking is a five-step cycle aimed at ensuring continuous improvement.

At the commencement of the process it would be appropriate to form a small benchmarking team or group, perhaps comprising a senior manager, health and safety specialist, line managers, employee representatives and representatives from the benchmarking partner or trade association.

Step 1 – Deciding what to benchmark

Benchmarking can be applied to any aspect of health and safety, but it is good practice to prioritise in terms of high hazard and risk areas, such as with the use of hazardous substances, with certain types of workplace or working practice. Feedback from safety monitoring activities, the risk assessment process and accident data should identify these priorities. Consultation with the workforce should take place at this stage, together with trade associations who may have experience of the process.

Step 2 – Deciding where you are

This stage of the exercise is concerned with identifying the current level of performance in the selected area for consideration and the desired improvement in performance. Reference should be made at this stage to legal requirements, such as regulations, to ACOPs and HSE guidance on the subject, and to any in-house statistical information. It may be appropriate to use an audit and/or questionnaire approach to measure the current level of performance.

Step 3 – Selecting partners

In large organisations it may be appropriate to select partners both from within the organisation, perhaps at a different geographical location (internal benchmarking) and from outside the organisation (external benchmarking). With smaller organisations, trade associations or the local Chamber of Commerce may be able to assist in the selection of partners. Local benchmarking clubs operate in a number of areas. Reference should be made to the *Benchmarking Code of Conduct* to ensure compliance with same at this stage.

Step 4 – Working with your partner
With the right planning and preparation, this stage should be straightforward. Any information that is exchanged should be comparable, confidentiality should be respected and all partners should have a good understanding of the partner's process, activities and business objectives.

Step 5 – Acting on lessons learned
Fundamentally, the outcome of any benchmarking exercise is to learn from other organisations, to learn more about the individual organisation's performance compared with the working partners and to take action to improve performance.

SMARTT
According to the HSE, any action plan should be 'SMARTT', that is:
- Specific
- Measurable
- Agreed
- Realistic
- Trackable **and**
- Timebound.

As with any action plan, it should identify a series of recommendations, the members of the organisation responsible for implementing these recommendations and a timescale for their implementation. Progress in implementation should be monitored on a regular basis. In some cases it may be necessary to redefine objectives in the light, for example, of recent legislation.

There should be a continuing liaison with benchmarking partners during the various stages of the action plan.

Pointers to success

To succeed in health and safety benchmarking, there should be:
- senior management resources and commitment
- employee involvement
- a commitment to an open and participatory approach to health and safety, including a willingness to share information with others within and outside the organisation
- comparison with data on a meaningful 'apples with apples' basis; and
- adequate research, planning and preparation.

BS 8800: Guide to occupational health and safety management systems

BS 8800: 2004 offers an organisation the opportunity to review and revise its current occupational health and safety arrangements against a standard that has been developed by industry, commerce, insurers, regulators, trade unions and occupational health and safety practitioners.

The aims of the standard are 'to improve the occupational health and safety performance of organisations by providing guidance of how management of occupational health and safety may be integrated with the management of other aspects of the business performance in order to:
- minimise risks to employees and others;
- improve business performance; and
- assist organisations to establish a responsible image in the workplace.'

In order to achieve positive benefits, health and safety management should be an integral feature of the undertaking contributing to the success of the organisation.

Status review

In any status review of the health and safety management system, BS 8800 recommends the following headings:

1. Requirements of relevant legislation dealing with health and safety management issues.
2. Existing guidance on health and safety management within the organisation.
3. Best practice and performance in the organisation's employment sector and other appropriate sectors e.g. from relevant HSC's industry advisory committees and trade association guidelines.
4. Efficiency and effectiveness of existing resources devoted to health and safety management.

Policies

BS 8800 identifies nine key areas that should be addressed in a policy, each of which allows visible objectives and targets to be set:

- recognising that occupational health and safety is an integral part of its business performance;
- achieving a high level of health and safety performance, with compliance to legal requirements as the minimum and continual cost effective improvement in performance;
- provision of adequate and appropriate resources to implement the policy;
- the publishing and setting of health and safety objectives, even if only by internal notification;
- placing the management of health and safety as a prime responsibility of line management, from most senior executive to first-line supervisory level;
- ensuring understanding, implementation and maintenance of the policy statement at all levels in the organisation;
- employee involvement and consultation to gain commitment to the policy and its implementation;

- periodic review of the policy, the management system and audit of compliance to policy;
- ensuring that employees at all levels receive appropriate training and are competent to carry out their duties and responsibilities.

The models
There are two recommended approaches depending upon the organisational needs of the business and with the objective that such an approach will be integrated into the total management system, namely:

(a) one based on *Successful health and safety management* [HS(G)65]; and
(b) one based on ISO 14001, which is compatible with the environmental standard.

Cleaning schedules

There is an implied duty in the Workplace (Health, Safety and Welfare) Regulations on employers to keep the workplace clean. Any management system for dealing with this matter should involve the use of formally written and supervised cleaning schedules. Compliance with cleaning schedules should be monitored on a frequent basis.

Elements of a cleaning schedule
A cleaning schedule should incorporate the following elements:

- the item or area to be cleaned
- the method, materials and equipment to be used
- the frequency of cleaning
- individual responsibility for ensuring the cleaning task is completed satisfactorily
- specific precautions necessary e.g. in the use of cleaning chemicals, segregation of areas to be cleaned.

Cleaning schedules should be laid out in tabular form incorporating the above elements.

Those involved in monitoring the effectiveness of the cleaning schedule should be trained in inspection techniques and should have sufficient authority within the organisation to require immediate action where failure to implement the parts of the schedule has occurred.

Competent persons

Health and safety legislation frequently requires an employer to appoint competent persons for a range of purposes. In general terms, 'competence' implies the possession of skill, knowledge and experience with respect to the tasks undertaken by that competent person.

According to *Brazier v Skipton Rock Company Limited (1962) 1 AER,* a competent person should have practical and theoretical knowledge as well as sufficient experience of the particular machinery, plant or procedure involved as will enable him to identify defects or weaknesses during plant or machinery examinations, and to assess their importance in relation to the strength and function of that plant and machinery.

The duty to appoint competent persons is covered in the Management of Health and Safety at Work Regulations 1999, thus:

Regulation 7 – Health and safety assistance
Every employer shall appoint one or more competent persons to assist him in undertaking the measures he needs to take to comply with the requirements and prohibitions imposed upon him by or under the relevant statutory provisions.

A person shall be regarded as competent where he has sufficient training and experience or knowledge and other qualities to enable him properly to assist the employer in undertaking the measures he needs to take to comply with the requirements and prohibitions imposed upon the employer by or under the relevant statutory provisions.

Regulation 8 – Procedures for serious and imminent danger and for danger areas
An employer must nominate a sufficient number of competent persons to implement those procedures as they relate to evacuation from premises of persons at work in his undertaking and ensure that none of his employees has access to any area occupied by him to which it is necessary to restrict access on the grounds of health and safety unless the employee concerned has received adequate health and safety instruction.

Further information is incorporated in the ACOP to the Regulations.

Other health and safety legislation requiring the appointment of competent persons includes:
Mines and Quarries Act 1954
Construction (Design and Management) Regulations 1994
Construction (Health, Safety and Welfare) Regulations 1996
Electricity at Work Regulations 1989
Ionising Radiations Regulations 1999
Lifting Operations and Lifting Equipment Regulations 1998
Control of Noise at Work Regulations 2005
Pressure Systems Safety Regulations 2000
Provision and Use of Work Equipment Regulations 1998
Work at Height Regulations 2005

 1(c) Principal regulations
Construction (Design and Management) Regulations 1994
Construction (Health, Safety and Welfare) Regulations 1996
Control of Noise at Work Regulations 2005
Electricity at Work Regulations 1989
Ionising Radiations Regulations 1999
Lifting Operations and Lifting Equipment Regulations 1998
Pressure Systems Safety Regulations 2000
Provision and Use of Work Equipment Regulations 1998
Work at Height Regulations 2005

Consequence analysis

Consequence analysis is a feature of risk analysis which considers the physical effects of a particular process failure and the damage caused by these effects. It is undertaken to form an opinion on potentially serious hazardous outcomes of accidents and their possible consequences for people and the environment. The technique should be used as a tool in the decision-making process in a safety study which incorporates the following features:

(a) description of the process system to be investigated;
(b) identification of the undesirable events;
(c) determination of the magnitude of the resulting physical effects;
(d) determination of the damage;
(e) estimation of the probability of the occurrence of calculated damage; and
(f) assessment of the risk against established criteria.

The outcome of consequence analysis is:

(a) for the chemical and process industries, to obtain information about all known and unknown effects that are of importance when something goes wrong in the plant and to obtain information on measures for dealing with catastrophic events;
(b) for the designing industries, to obtain information on how to minimise the consequences of accidents;
(c) for the operators in the processing plant and people living in the immediate vicinity, to give them an understanding of their personal situation and the measures being taken to protect them; and
(d) for the enforcement and legislative authorities, to provide them with information on measures being taken to ensure compliance with current legal requirements.

Consequence analysis is generally undertaken by a team of specialists, including chemists and safety technologists, experienced in the actual problems of the system concerned.

Dose record (ionising radiation)

Where employees may be exposed to ionising radiation, an individual dose record must be maintained by their employer.

In relation to a person, 'dose record' means the record of the doses received by that person as a result of his exposure to ionising radiation, being the record made and maintained on behalf of the employer by the approved dosimetry service in accordance with Regulation 21 of the Ionising Radiations Regulations.

Records must be made or maintained until that person has or would have attained the age of 75 years but in any event for at least 50 years from when they were made.

 1(c) Principal regulations
    Ionising Radiations Regulations 1999

Event tree analysis

This technique is similar to fault tree analysis, working from a selected 'initiating event', such as an electrical fault in a manufacturing system. It is, basically, a systematic representation of all the possible states of the processing system conditional to the specific initiating event and relevant for a certain type of outcome, such as a major fire or unsafe feature of the manufacturing system.

Failure mode and effect analysis

This technique is based on identifying the possible failure modes of each component of a system and predicting the consequences of that failure. For example, if a safety device linked to a machinery guard fails, it could result in the operator being exposed to danger. As a result, attention is paid to those consequences at the design stage of the machinery safety system and in the preparation of the planned preventive maintenance procedure for the machine.

Fault tree analysis

A form of safety management technique which begins with the consideration of a chosen 'top event', such as a pressure vessel explosion, and then assesses the combination of failures and conditions which could cause this event to take place.

This technique is used widely in quantitative risk analysis, particularly where control over process controls is critical to meeting safety standards.

Health and safety file

This is a document containing information for the client or user of a building on the risks that may be present during maintenance, repair or renovation. Under the Construction (Design and Management) Regulations, the appointed planning supervisor for a project must ensure that the health and safety file is prepared and delivered to the client at the end of the project.

The client must take reasonable steps to ensure that the information in any health and safety file is kept available for inspection by any person who may need information in the file for the purpose of complying with the requirements and prohibitions imposed upon him by or under the relevant statutory provisions.

 1(c) Principal regulations
   Construction (Design and Management) Regulations 1994
 1(d) Approved code of practice
   Managing construction for health and safety

Health and safety plans

Requirements relating to the preparation of health and safety plans for a project are covered in Regulation 15 of the Construction (Design and Management) Regulations.

Pre-tender stage health and safety plan
A planning supervisor appointed for a project must ensure that a health and safety plan has been prepared prior to the project commencing.

Construction phase health and safety plan
The principal contractor must take such measures as is reasonable to ensure that a health and safety plan is prepared which covers the construction phase.

The contents of both types of health and safety plan are specified in the regulations and ACOP.

 1(c) Principal regulations
   Construction (Design and Management) Regulations 1994
 1(d) Approved codes of practice
   Managing construction for health and safety

Health and safety training

The Health and Safety at Work etc. Act, together with many regulations, place either absolute or qualified duties on employers to provide health and safety training for employees.

As with any training process, establishment of training programmes should follow a series of clearly defined stages:
- Identification of training needs:
    - What kind of training is required?
    - When will the training be needed?
    - How many people need to be trained?

  ○ What is the standard of performance required of trainees following the training?
- development of the training plan and programme:
  ○ What are the training objectives?
  ○ What has to be taught – theory and practice?
  ○ What is the best method of teaching?
- Implementation of the training programme:
  ○ organising the training
  ○ undertaking the training
  ○ recording the results
  ○ evaluation of the results.

The need for health and safety training may be identified as an outcome of the risk assessment process, safety monitoring activities, such as safety audits, and the investigation of accidents, ill-health and incidents.

 1(b) The principal statutes
Health and Safety at Work etc. Act 1974

 1(c) Principal regulations
Construction (Health, Safety and Welfare) Regulations 1996
Control of Asbestos at Work Regulations 2002
Control of Lead at Work Regulations 2002
Control of Substances Hazardous to Health Regulations 2002
Health and Safety (Display Screen Equipment) Regulations 1992
Ionising Radiations Regulations 1999
Management of Health and Safety at Work Regulations 1999
Personal Protective Equipment at Work Regulations 1992
Provision and Use of Work Equipment Regulations 1998

Health records

The maintenance of individual health records is a standard feature of occupational health practice. The purpose of such records is to:

- assist occupational health practitioners to provide efficient health surveillance, emergency attention, health care and continuity of such care;

- enable practitioners to undertake epidemiological studies to identify general health and safety risks and trends arising amongst employees and to identify problem areas and specific risks;
- establish, maintain and keep up-to-date written information relating to people, hazards and current monitoring activities; and
- facilitate assessment of problems, decision making, recommendations and the writing of reports.

1(c) Principal regulations
Control of Asbestos at Work Regulations 2002
Control of Substances Hazardous to Health Regulations 2002
Ionising Radiations Regulations 1999

Health surveillance

Health surveillance implies the specific health examination at a pre-determined frequency of those at risk of developing further ill-health or disability and those actually or potentially at risk by virtue of the type of work they undertake during their employment.

A risk assessment will identify the circumstances in which health surveillance is required by specific health and safety regulations, such as the Control of Substances Hazardous to Health (COSHH) Regulations. Health surveillance should also be introduced where a risk assessment shows that the following criteria apply:

(a) there is an identifiable disease or adverse health condition related to the work concerned;
(b) valid techniques are available to detect indications of the disease or condition;
(c) there is a reasonable likelihood that the disease or condition may occur under the particular conditions of work; and
(d) surveillance is likely to further the protection of the health and safety of the employees to be covered.

The appropriate level, frequency and procedure of health surveillance should be determined by a competent person (e.g. occupational health nurse) acting within the limits of their training and experience. This could be determined on the basis of suitable general guidance (e.g. regarding skin inspection for dermal effects) but, in certain circumstances, this may require the assistance of a qualified medical practitioner. The minimum requirement for health surveillance is keeping a health record. Once it is decided that health surveillance is appropriate, it should be maintained through an employee's employment unless the risk to which the worker is exposed and associated health effects are rare and short term.

[ACOP to the Management of Health and Safety at Work Regulations]

 1(c) Principal regulations

Control of Asbestos at Work Regulations 2002
Control of Lead at Work Regulations 2002
Control of Noise at Work Regulations 2005
Control of Substances Hazardous to Health Regulations 2002
Control of Vibration at Work Regulations 2005
Health and Safety (display Screen Equipment) Regulations 1992
Ionising Radiations Regulations 1999
Management of Health and Safety at Work Regulations 1999

 3(b) Forms

Occupational Health
(a) Pre-employment health questionnaire
(b) Health questionnaire
(c) Food handler's clearance certificate
(d) Fitness certificate

 Appendix B: Documentation and record keeping requirements

Information and instruction

The provision of health and safety information and instruction for employees is a common requirement of health and safety legislation, including the Health and Safety at Work etc. Act 1974 and regulations, such as the Health and Safety (Display Screen Equipment) Regulations 1992, the Personal Protective Equipment at Work Regulations 1992 and the Noise at Work Regulations 1989.

 1(b) The principal statutes
   Health and Safety at Work etc. Act 1974
 1(c) Principal regulations
   Control of Asbestos at Work Regulations 2002
   Control of Lead at Work Regulations 2002
   Control of Substances Hazardous to Health Regulations 2002
   Health and Safety (Display Screen Equipment) Regulations 1992
   Ionising Radiations Regulations 1999
   Management of Health and Safety at Work Regulations 1999
   Noise at Work Regulations 1999
   Personal Protective Equipment at Work Regulations 1992
   Safety Representatives and Safety Committees Regulations 1977

International Loss Control Institute loss causation model

The International Loss Control Institute (ILCI) views the safety management process as being concerned, fundamentally, with the prevention of loss.

Lack of health and safety control within an organisation, such as a failure to undertake planned inspections and set down organisational rules, creates the basic causes for accidents and occupational ill-health. These basic causes may be associated with both personal factors, such as stress, and job factors, such as inadequate maintenance of work equipment.

The basic causes contribute to the immediate, or direct, causes associated with sub-standard practices and conditions. Sub-standard practices include, for example, the removal of safety devices from machinery. Sub-standard conditions, on the other hand, may be associated with excessive noise or inadequate ventilation in a workplace.

In turn, the immediate or direct causes create the incident, such as a major injury arising from a fall or contact with dangerous parts of machinery.

1. Lack of control

Failure to maintain compliance with adequate standards for:
- Leadership and administration
- Management training
- Planned inspections
- Job task analysis and procedures
- Job and task observations
- Job task observations
- Emergency preparedness
- Organisational rules
- Accident and incident investigations
- Accident and incident analysis
- Personal protective equipment
- Health control and services
- Programme evaluations systems
- Purchasing and engineering systems
- Personal communications
- Group meetings
- General promotion
- Hiring and placement
- Records and reports
- Off-the-job safety

CREATES

2. Basic causes

Personal factors
- Inadequate capability
  - physical/physiological
  - mental/psychological
- Lack of knowledge
- Lack of skill
- Stress
  - physical/physiological
  - mental/psychological

Job factors
- Inadequate leadership or supervision
- Inadequate engineering
- Inadequate purchasing
- Inadequate maintenance
- Inadequate tools, equipment, materials
- Inadequate work standards
- Abuse and misuse
- Wear and tear

LEADING TO:

3. Immediate causes

Substandard practices
- Operating equipment without authority
- Failure to warn
- Failure to secure
- Operating at improper speed
- Making safety devices inoperable
- Removing safety devices
- Using defective equipment
- Failing to use personal protective equipment properly
- Improper loading
- Improper placement
- Improper lifting
- Improper position for task
- Servicing equipment in operation
- Horseplay
- Under influence of alcohol/drugs

Substandard conditions
- Inadequate guards or barriers
- Inadequate or improper protective equipment
- Defective tools, equipment, materials
- Congestion or restricted action
- Inadequate warning system
- Fire and explosion hazards
- Poor housekeeping, disorder
- Noise exposure
- Radiation exposure
- Temperature extremes
- Inadequate or excess illumination
- Inadequate ventilation

CAUSING:

4. Incident

Contacts
- Struck against
- Fall to lower level
- Caught in
- Caught between
- Overstress, overexertion, overload
- Struck by
- Fall on same level
- Caught on
- Contact with

WITH THE RESULTING:

5. Loss

Personal harm
- Major injury or illness
- Serious injury or illness
- Minor injury or illness

Property Damage
- Catastrophic
- Major
- Serious
- Minor

Process loss
- Catastrophic
- Major
- Serious
- Minor

The ILCI loss causation model

The incident results in loss, both to the accident victim and to the organisation.

The theory behind this model is that, by concentrating on good standards of health and safety management and control, the indirect and direct causes of accidents can be greatly reduced leading to a comparative reduction in loss-producing incidents. Particular attention must be paid to people in terms of their individual skills and knowledge, the jobs they carry out, working practices and work conditions.

ISO 14001: Environmental Management Systems

This Standard provides a model for health and safety management systems. Implementation of the Standard takes place in a number of clearly defined stages, thus:

1. Initial status review
This stage entails a review and assessment of the current 'state of play' with regard to health and safety management systems. Proactive factors to be considered include the presence of written safe systems of work, joint consultation procedures, an integrated approach to risk assessment, documented planned preventive maintenance systems and a procedure for providing information, instruction and training at all levels within the organisation.

Reactive management systems include those for accident and incident reporting, recording and investigation, accident and incident costing and means for the provision of feedback following the investigation of accidents, incidents and occupational ill-health.

2. Occupational health and safety policy
A review of the current Statement of Health and Safety Policy and other sub-policies covering, for example, stress at work,

contractors' activities and the provision of personal protective equipment, takes place at this stage.

3. Planning
Feedback from the initial status review and assessment of the effectiveness of the Statement of Health and Safety Policy will identify areas for planning for future actions. This stage may entail the establishment of management systems to cover:
(a) future safety monitoring operations;
(b) the preparation of rules for the safe conduct of project work (contractors' regulations)
(c) systems for raising the awareness of employees;
(d) the provision of information, instruction and training;
(e) planned preventive maintenance;
(f) health surveillance of specific groups of employees; and
(g) a review of risk assessment procedures.

4. Implementation and operation
Once the strategies and objectives for future health and safety activities have been established at the planning stage, the process of implementing these objectives must be put into operation, perhaps on a phased basis. The written objectives should specify:
(a) the actual objective;
(b) the manager responsible for achieving this objective;
(c) the financial arrangements where appropriate;
(d) the criteria for assessing successful achievement of the objective; and
(e) a date for completion of the objective.

5. Checking and corrective action
Procedures should be established for ensuring that agreed objectives are being achieved within the timescale allocated and for ensuring specific corrective action is taken in the event of failure or incomplete fulfilment of the objective.

6. Management review

Any phased programme of improvement must be subject to regular management review. The timescale for review, and the management responsible for same, should be established before the implementation stage. In most cases a review team would assess the success in achievement of the pre-determined objectives and make recommendations for future action, including any safety monitoring arrangements necessary.

7. Continual improvement

As a result of undertaking this phased approach to health and safety management, there should be continual improvement in health and safety performance including:

(a) improved attitudes and awareness on the part of management and employees;

(b) greater commitment to, and recognition of, the need to incorporate health and safety in management procedures;

(c) regular revisions of policy based on feedback from reviews;

(d) a developing health and safety culture within the organisation;

(e) improved systems for ensuring corrective action is dealt with quickly; and

(f) ease of integration of environmental management systems with health and safety management systems.

Joint consultation

Joint consultation is an important means of improving motivation of employees and others by enabling them to participate in planning work and setting objectives. The process of consulting on health and safety issues, procedures and systems may take place through discussions by an employer with trade-union-appointed safety representatives, non-trade-union representatives of employee safety, and through the operation of a safety committee.

There is an absolute duty on an employer under the Health and Safety at Work etc. Act 1974 (Section 2(6)) to consult with safety representatives with a view to the making and maintenance of arrangements which will enable him and his employees to co-operate effectively in promoting and developing measures to ensure the health and safety at work of the employees, and in checking the effectiveness of such measures.

Further legal and practical requirements relating to joint consultation are laid down in the Safety Representatives and Safety Committees Regulations 1977 and the Health and Safety (Consultation with Employees) Regulations 1996, together with accompanying ACOP and HSE Guidance.

 1(b) The principal statutes
Health and Safety at Work etc. Act 1974

1(c) Principal regulations
Health and Safety (Consultation with Employees) Regulations 1996
Safety Representatives and Safety Committees Regulations 1977

 1(d) Approved codes of practice
Safety representatives and safety committees

Local rules

Under the Ionising Radiations Regulations, every radiation employer (as defined) shall, in respect of any controlled area and supervised area, make and set down in writing such local rules as are appropriate to the radiation risk and the nature of the operations undertaken in that area (Regulation 17).

A radiation employer shall take all reasonable steps to ensure local rules are observed and brought to the attention of appropriate employees and other persons.

The radiation employer shall appoint one or more radiation protection supervisors to ensure compliance with the Regulations in respect of any area made subject to local rules.

1(c) Principal regulations
Ionising Radiations Regulations 1999

 1(d) Approved codes of practice
Work with ionising radiation

Major incidents

A major incident is one that may:
- affect several locations or departments within a workplace, e.g. a major escalating fire;
- endanger the surrounding communities, such as a pollution incident;
- be classed as a fatal or major injury accident, or a scheduled dangerous occurrence, under the Reporting of Injuries, Diseases and Dangerous Occurrences Regulations; or
- result in adverse publicity for an organisation with ensuing loss of public confidence and market place image, e.g. a product recall.

Organisations need to have a formal major incident policy together with established procedures covering the principal stages of an emergency situation. A typical emergency procedure covers the following stages:
- preliminary action;
- action when emergency is imminent;
- action during the emergency;
- ending the emergency.

Implementation of the emergency procedure would normally incorporate:
- liaison with external authorities, such as the HSE, fire authority, local authority;
- the appointment of an emergency controller and establishment of an emergency control centre;
- nomination of senior managers responsible for initiating the procedure;
- notification to local authorities;
- call-out of designated competent persons, e.g. engineers, health and safety personnel;

- immediate action on site (by supervisors and employees);
- evacuation procedure and nomination of competent persons to oversee same;
- access to records of employees;
- external communication arrangements;
- public relations;
- catering and temporary shelter arrangements;
- contingency arrangements covering repairs to buildings, etc.; and
- training exercises involving external services, such as the fire brigade and ambulance service.

 1(c) Principal regulations
Ionising Radiations Regulations 1999
Management of Health and Safety at Work Regulations 1999

 1(d) Approved codes of practice
Management of health and safety at work
Prevention of fire and explosion and emergency response on offshore installations
Work with ionising radiation

Method statements

This is a form of written safe system of work commonly used in construction activities where work with a foreseeably high hazard content is to be undertaken. The system of work may be agreed between an occupier and a principal contractor or between a principal contractor and sub-contractor.

A method statement should specify, on a stage-by-stage basis, the operations to be undertaken and the precautions necessary to protect all persons on site, members of the public and local residents. It may incorporate information and specific requirements stipulated by, for example, health and safety specialists, enforcement officers, site surveyors, police and manufacturers of articles and substances used in the work. In certain cases it may identify training needs or the use of specifically trained operators.

Typical situations requiring the production of a method statement are:
- the use of substances hazardous to health;
- the use of explosives;
- lifting operations on site;
- potential fire hazard situations;
- where electrical hazards may raise;
- the use of sealed sources of radiation;
- excavation adjacent to existing buildings;
- demolition activities; and
- work involving asbestos removal or stripping.

A method statement should incorporate the following elements:
- the technique(s) to be used for the work;
- access provisions;
- procedures for safeguarding existing locations;
- structural stability precautions;
- procedures for ensuring the safety of others, including members of the public and local residents;
- health precautions, including the use of certain forms of personal protective equipment, such as respiratory protection;
- the plant and equipment to be used;
- procedures for the prevention of area pollution;
- procedures for segregating certain areas;
- procedures for disposal of hazardous substances; and
- procedures for ensuring compliance with specific Regulations, such as the Control of Lead at Work Regulations, Control of Asbestos at Work Regulations.

☞ **1(c) Principal regulations**
   Construction (Design and Management) Regulations 1994
   Construction (Health, Safety and Welfare) Regulations 1996
☞ **1(d) Approved codes of practice**
   Managing construction for health and safety
☞ **1(e) HSE guidance notes**
   Health and safety in construction [HS(G)150]

Management oversight and risk tree (MORT)

MORT is defined as 'a systemic approach to the management of risks in an organisation'. It was developed by the United States Department of Energy during the period 1978–83, and incorporates methods aimed at increasing reliability, assessing the risks, controlling losses and allocating resources effectively.

The MORT philosophy is summarised under the following four headings.

Management takes risks of many kinds
Specifically, these risks are classified in the areas of:
   (a) product quantity and quality;
   (b) cost;
   (c) schedule;
   (d) environment, health and safety.

Risks in one area affect operations in other areas
Management's job may be viewed as one of balancing risks. For instance, to focus only on safety and environmental issues would increase the risk of losses from deficiencies, schedule delays and costs.

Risks should be made explicit where practicable
Since management must take risks, it should know the potential consequences of those risks.

Risk management tools should be flexible enough to suit a variety of diverse situations
While some analytical tools are needed for complex situations, other situations require simpler and quicker approaches. The MORT system is designed to be applied to all of an organisation's risk management concerns, from simple to complex.

The MORT process

The acronym, MORT, carries two primary meanings:

   (a) the MORT 'tree' or logic diagram, which organises risk, loss and safety programme elements and is used as a master worksheet for accident investigations and programme evaluations; and

   (b) the total safety programme, seen as a sub-system to the major management system of an organisation.

The MORT process includes four main analytical tools as follows.

Change analysis

This is based on the Kepner-Tregoe method of rational decision-making. Change analysis compares a problem-free situation with a problem (accident) situation in order to isolate causes and effects of change. It is especially useful when the decision-maker needs a quick analysis, when the cause is obscure, and when well-behaved personnel behave differently from past situations, as with the Three Mile Island incident.

Energy trace and barrier analysis (ETBA)

ETBA is based on the notion that energy is necessary to do work, that energy must be controlled, and that uncontrolled energy flowing in the absence of adequate barriers can cause accidents. The simple 'energy-barrier-targets' concept is expanded with the details of specific situations to answer the question '*What happened?*' in an accident. ETBA may be performed very quickly or applied meticulously as time permits.

MORT tree analysis

This is the third and most complex tool, combining principles from the fields of management and safety. It uses fault tree methodology with a view to assisting the investigator to ascertain what happened and why it happened. The MORT tree organises over 1500 basic events (causes) leading to 98 generic events (problems). Both specific control factors and management system factors are analysed for their contributions to the accident. People, procedures and hardware are considered separately, and then together, as key system safety elements.

Positive (success) tree design
This technique reverses the logic of fault tree analysis. In positive tree design, a system for successful operation is comprehensively and logically laid out. The positive tree is an excellent planning and assessment tool because it shows all that must be performed and the proper sequencing of events needed to accomplish an objective.

Objectives of the MORT technique
MORT is, fundamentally, an analytical technique or procedure to determine the potential for downgrading incidents in situations. It places special emphasis on the part that management oversight plays in allowing untoward or adverse events to occur. The MORT system is designed to:

(a) result in a reduction in oversights, whether by omission or commission, that could lead to downgrading incidents if they are not corrected;

(b) determine the order of risks and refer them to the proper organisational level for corrective action;

(c) ensure best allocation and use of resources to organise efforts to prevent or reduce the number and severity of adverse incidents.

OHSAS 18001: A Pro-active Approach to Health and Safety Management

This standard specifies a staged approach for developing and implementing a plan, incorporating key stages (Refer to Flowchart given in the next page).

Planned preventive maintenance

Under the Management of Health and Safety at Work Regulations there is an implied duty to manage those workplaces and

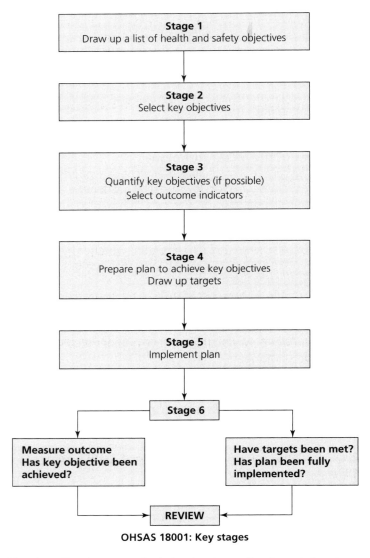

OHSAS 18001: Key stages

The above flowchart covers both the planning and implementation stages to indicate the complete process. Planning involves Stages 1 to 4.

activities which could result in the creation of risks. More specifically, the duty on employers to maintain the workplace and work equipment are incorporated in regulations as follows.

Workplace (Health, Safety and Welfare) Regulations 1992

Regulation 5 – Maintenance of workplace, and of equipment, devices and systems

The workplace and the equipment, devices and systems to which this regulation applies shall be maintained (including cleaned as appropriate) in an efficient state, in efficient working order and in good repair.

Where appropriate, the equipment, devices and systems to which this regulation applies shall be subject to a suitable system of maintenance.

The equipment, devices and systems to which this regulation applies are:
  (a) equipment and devices a fault in which is liable to result in a failure to comply with any of these regulations; and
  (b) mechanical ventilation systems provided pursuant to Regulation 6 (whether or not they include equipment or devices within sub-paragraph (a) of this paragraph).

Provision and Use of Work Equipment Regulations 1998

Regulation 5 – Maintenance

Every employer shall ensure that work equipment is maintained in an efficient state, in efficient working order and in good repair.

Every employer shall ensure that where any machinery has a maintenance log, the log is kept up to date.

Features of a planned preventive maintenance system

A formally written system should incorporate the following elements:

- the structural item, area, system e.g. floor surface, ventilation system, device such as a safety device on a machine, or item of machinery to be maintained;
- the maintenance procedure to be followed;
- the frequency of maintenance;
- individual management responsibility for ensuring the maintenance procedure is implemented; and
- specific precautions necessary, e.g. operation of a permit-to-work system, isolation of the area, display of signs and notices, and restriction of certain work to designated employees who have been trained in the maintenance procedure.

A planned preventive maintenance system should be produced in tabular form incorporating the above elements. Management should be able to assess, at any point in time, progress is otherwise in the implementation of the system.

 1(c) Principal regulations
Provision and Use of Work Equipment Regulations 1998
Workplace (Health, Safety and Welfare) Regulations 1992

 1(d) Approved codes of practice
Safe use of work equipment
Safe use of power presses
Safe use of woodworking machinery

Quality systems audit (QSA)

This formal audit system follows the principles of the International Standards Organisation for Quality Systems Auditing (ISO 19011). Evidence of compliance is gained by way of examination of documentation, the questioning of employees at all levels, and observation of physical conditions at the workplace.

QSA is divided into five main sections and eleven sub-sections, corresponding to the structure of the HSE Guidance Note

HS(G)65 *Successful health and safety management*. QSA takes a structured approach to examining an organisation's health and safety management system and includes all the elements of OHSAS 18001 (see above). Guidance for auditors is incorporated in the audit workbook for the system.

Scoring

QSA uses an 'all or nothing' approach, that is, all the points can be awarded in respect of a particular question or none at all, in the later case being directed at encouraging improvement in performance. Non-applicable question scores can be deducted from the total potential.

Award system

There is an award system based on evaluation of all sub-sections of the audit in all cases, consisting of five possible levels based on the minimum sub-section percentage score. The lower number of audit sections and sub-sections under QSA allows the use of the 'radar chart', a powerful mapping tool for demonstrating at a glance the strengths and weaknesses of the organisation's health and safety management system.

Training

Formal training is necessary for people using the system, providing the options for 'internal' audits using the organisation's own personnel or 'external' audits using the owners of the audit system. In the case of QSA, the system is owned by the Royal Society for the Prevention of Accidents (RoSPA).

Risk assessment

Risk assessment is the principal feature of all modern protective legislation – health and safety, food safety and environmental protection.

The Management of Health and Safety at Work Regulations place an absolute duty on every employer to make a suitable and sufficient assessment of:

(a) the risks to the health and safety of employees to which they are exposed whilst at work; and

(b) the risks to the health and safety of persons not in their employment arising out of or in connection with the conduct by him of his undertaking;

for the purpose of identifying the measures he needs to take to comply with the requirements and prohibitions imposed upon him by or under the relevant statutory provisions.

A 'suitable and sufficient risk assessment' should:

(a) identify the significant risks arising out of the work;

(b) enable the employer to identify and prioritise the measures that need to be taken to comply with the relevant statutory provisions; and

(c) be appropriate to the nature of the work and such that it remains in force for a reasonable period of time.

Further information on risk assessment is incorporated in Regulation 3 of the regulations and the ACOP to same.

 1(c) Principal regulations

Control of Asbestos at Work Regulations 2002
Control of Lead at Work Regulations 2002
Control of Noise at Work Regulations 2005
Control of Substances Hazardous to Health Regulations 2002
Control of Vibration at Work Regulations 2005
Dangerous Substances and Explosive Atmospheres Regulations 2002
Health and Safety (Display Screen Equipment) Regulations 1992
Ionising Radiations Regulations 1999
Management of Health and Safety at Work Regulations 1999
Manual Handling Operations Regulations 1992
Personal Protective Equipment at Work Regulations 1992
Work at Height Regulations 2005
Regulatory Reform (Fire Safety) Order 2005

☞ **1(d) Approved codes of practice**

Control of asbestos at work
Control of lead at work
The management of asbestos in non-domestic premises
Control of substances hazardous to health
Dangerous substances and explosive atmospheres
Work with ionising radiation
Management of health and safety at work

☞ **1(e) HSE guidance notes**

*Assessing and managing risks at work from skin exposure to
   chemical agents*
A step-by-step guide to COSHH assessment
Five steps to risk assessment

Risk management

Risk management is variously defined as:
- the minimisation of the adverse effects of pure and speculative risks within a business;
- the identification, measurement and economic control of
  the risks that threaten the assets and earnings of a business or other enterprise;
- the identification and evaluation of risk and the determination of the best financial solution for coping with the
  major and minor threats to a company's earnings and
  performance;
- a technique for coping with the effects of change.

Risk management techniques have the principal objective of
producing savings in insurance premiums by first defining and
then minimising areas of industrial and other risk. It seeks not
to discredit insurance arrangements but to promote the concept of insuring only what is necessary in terms of risk. On this
basis the manageable risks are identified, measured and either
eliminated or controlled, and the financing of the remaining or
residual risks, normally through insurance, takes place at a later
stage.

Categories of risk

There are two main areas of risk, namely catastrophic risk, which demands insurance, and risks associated with wastage of the organisation's assets. The latter is where the scope of self-insurance and diminution of risk is most evident, and is why organisations appoint risk managers, in some cases establishing risk management subsidiaries.

Risks may be of a pure or speculative nature. *Pure risks* can only result in loss to the organisation. *Speculative risks*, on the other hand, may result in either gain or loss. Within the context of a risk management programme, risk may be defined as 'the chance of loss', and the programme is therefore geared to safeguarding the organisation's assets, namely manpower, materials, machinery, methods, manufactured goods and money.

The risk management process

This takes place in a series of stages:

  (a) identification of the exposure to risk, such as that arising from fire, storm and flood, accidents, human error, theft or fraud, breach of legislation, etc.;
  (b) analysis and evaluation of the identified exposures to risk;
  (c) risk control, using a range of protective measures; and
  (d) financing of the risk at the lowest cost.

Risk control strategies

Risk avoidance	Risk retention	Risk transfer
This strategy involves a conscious decision on the part of the organisation to avoid completely a risk by discontinuing the operation or circumstances that produces the risk.	In this case, the risk is retained within the organisation where any consequent loss is financed by that organisation.	This is the legal assignment of the costs of certain potential losses from one party to another, e.g. from a company to an insurance company.

Safe systems of work

This may be defined as the integration of people, machinery and materials in a safe and healthy environment and workplace to produce and maintain an acceptable standard of safety.

Requirements for a safe system of work include:
- a safe workplace layout with adequate space;
- a safe means of access to and egress from the working area;
- a correct sequence of operations;
- analysis of jobs, using techniques such as job analysis and job safety analysis;
- identification of safe procedures, both routine and emergency;
- a safe and healthy working environment in terms of temperature, lighting, ventilation and humidity, noise and vibration control, and hazardous airborne contaminants; and
- the provision of information, instruction, training and supervision for employees operating the system of work.

 1(b) Statutes
Health and Safety at Work etc. Act 1974

Safety monitoring systems

Active monitoring of the workplace and work activities should be undertaken through a range of techniques. These include the following.

Safety inspections
A scheduled or unscheduled inspection of a workplace to examine current levels of safety performance, working practices and compliance with legal requirements at a particular point in time. One of the principal objectives is the identification of hazards

and the making of recommendations, short, medium and long-term, to prevent or control exposure to these hazards.

Safety audits
The systematic measurement and validation of an organisation's management of its health and safety programme against a series of specific and attainable standards (Royal Society for the Prevention of Accidents).

A safety audit subjects each area of an organisation's activities to a systematic critical examination with the principal objective of minimising loss. It is an on-going process aimed at ensuring effective health and safety management.

Safety surveys
A detailed examination of a number of critical areas of operation or an in-depth study of the whole health and safety operation of premises.

Safety sampling exercises
An organised system of regular random sampling, the purpose of which is to obtain a measure of safety attitudes and possible sources of accidents by the systematic recording of hazard situations observed during inspections made along a predetermined route in a workplace.

Hazard and operability studies (HAZOPS)
These studies incorporate the application of formal critical examination to the process and engineering intentions for new facilities, such as production processes. The aim of HAZOPS is to assess the hazard potential arising from incorrect operation of each item of equipment and the consequential effects on the facility as a whole. Remedial action is then usually possible at a very early stage of the project with maximum effectiveness and minimum cost.

1(e) HSE guidance notes
Successful health and safety management

Safety signs

A 'safety sign' is defined as a sign that gives a message about health and safety by a combination of geometric form, safety colour and symbol or text, or both. The Safety Signs Regulations require that any sign displayed in a workplace must comply with the specification of signs contained in BS 5378: Part 1: 1980 *Safety Signs and Colours: Specifications for Colour and Design*. There are four basic categories of safety sign.

Prohibition
These signs indicate that certain behaviour is prohibited or must stop immediately, for example, smoking in a non-smoking area. These signs are recognised by a red circle with a cross running from top left to bottom right on a white background. Any symbol is reproduced in black within the circle.

Warning
These are signs which give warning or notice of a hazard. The signs are black outlined triangles filled in by the safety colour, yellow. The symbol or text is in black. The combination of black and yellow identifies the need for caution.

Mandatory
These signs indicate that a specific course of action is required, for example, ear protection must be worn. The safety colour is blue with the symbol or text in white. The sign is circular in shape.

Safe condition
The signs provide information about safe conditions. The signs are rectangular or square in shape, coloured green with white text or symbol.

Health and Safety (Safety Signs and Signals) Regulations 1996

These regulations cover various means of communicating health and safety information, including the use of illuminated signs, hand and acoustic signals (e.g. fire alarms), spoken communication and the marking of pipework containing dangerous substances.

Employers must use a safety sign where a risk cannot be adequately avoided or controlled by other means. The regulations require, where necessary, the use of road traffic signs within workplaces to regulate road traffic. Employers are required, firstly, to maintain the safety signs which are provided by them and, secondly, explain unfamiliar signs to their employees and tell them what they need to do when they see a safety sign.

The regulations also deal with fire safety signs including the need for exit signs to incorporate the Running Man symbol.

 1(d) HSE guidance notes

Safety signs and signals: Health and Safety (Safety Signs and Signals) Regulations 1996

Statements of health and safety policy

As stated in Part 1: Legal Background, a Statement of Health and Safety Policy incorporates three main elements.

Part 1: *General statement of intent*	Part 2: *Organisation*	Part 3: *Arrangements*
This part outlines the organisation's philosophy and objectives with respect to health and safety and should incorporate the duties of employers specified in Section 2 of the Health and Safety at Work etc. Act 1974.	It is useful to incorporate an organisational chart, or description of the chain of command, from the chief executive, managing director, senior partner, etc. downwards. This part should indicate clearly individual levels of responsibility and how accountability is fixed, the system for monitoring implementation of the policy and the relationship of the safety adviser with senior management.	Part 3 deals with the management systems and procedures which assist in overall policy implementation. It covers a wide range of matters including:

- the arrangements for risk assessment
- the arrangements for safe systems of work, including permit-to-work systems
- safety monitoring
- accident reporting, recording and investigation
- provision of information, instruction, training and supervision
- consultation with safety representatives and employees generally
- control of exposure to substances hazardous to health, noise, radiation, etc
- emergency procedures
- occupational health procedures
- fire safety arrangements, etc.

Appendices

It is common to incorporate a series of appendices such as:

- relevant statutory provisions applying to the organisation
- duties and responsibilities of all levels, plus the responsibilities of the safety adviser
- statements of policy on:
  - smoking at work
  - stress at work
  - health and safety training etc.
- the hazards arising and the precautions necessary on the part of all persons at work
- sources of health and safety information
- role and function of the safety committee.

 1(b) The principal statutes

Health and Safety at Work etc. Act 1974

 1(e) HSE guidance notes

Writing your health and safety policy statement: a guide to preparing a safety policy for a small business

Successful health and safety management [HS(G)65]

This HSE publication specifies elements of successful health and safety management within five main areas.

Policy

Organisations which are successful in achieving high standards of health and safety have health and safety policies which contribute to their business performance, while meeting their responsibilities to people and the environment in a way which fulfils both the spirit and the letter of the law. In this way they satisfy the expectations of shareholders, employees, customers and society at large. Their policies are cost-effective and aimed at achieving the preservation and development of physical and human resources and reductions in financial losses and liabilities. Their health and safety policies influence all their activities and decisions, including those to do with the selection of resources and information, the design and operation of working systems, the design and delivery of products and services, and the control and disposal of waste.

Organising

Organisations which achieve high health and safety standards are structured and operated so as to put their health and safety policies into effective practice. This is helped by the creation of a positive culture which secures involvement and participation at all levels. It is sustained by effective communication and the promotion of competence which enables all employees to make a responsible and informed contribution to the health and safety effort. The visible and active leadership of senior managers is necessary to develop and maintain a culture supportive of health and safety management. Their aim is not simply to avoid accidents, but to motivate and empower people to work safely. The visions, values and beliefs of leaders become the shared 'common knowledge' of all.

Planning

These successful organisations adopt a planned and systematic approach to policy implementation. Their aim is to minimise the risks created by work activities, products and services. They use risk assessment methods to decide priorities and set objectives for hazard elimination and risk reduction. Performance standards are established and performance is measured against them. Specific actions needed to promote a positive health and safety culture and to eliminate and control risks are identified. Wherever possible, risks are eliminated by the careful selection and design of facilities, equipment and processes or minimised by the use of physical control measures. Where this is not possible, systems of work and personal protective equipment are used to control risks.

Measuring performance

Health and safety performance in organisations which manage health and safety successfully is measured against pre-determined standards. This reveals when and where action is needed to improve performance. The success of action taken to control risks is assessed through active self-monitoring involving a range of techniques. This includes an examination of both hardware (premises, plant and substances) and software (people, procedures and systems), including individual behaviour. Failures of control are assessed through reactive monitoring which requires the thorough investigation of accidents, ill-health and incidents with the potential to cause harm or loss. In both active and reactive monitoring the objectives are not only to determine the immediate causes of sub-standard performance but, more importantly, to identify the underlying causes and the implications for the design and operation of the health and safety management systems.

Auditing and reviewing performance

Learning from **all** relevant experience and applying the lessons learned are important elements in effective health and safety management. This needs to be done systematically through regular reviews of performance based on data both from monitoring activities and from independent audits of the whole health and safety management system. These form the basis of self-regulation and for securing compliance with Sections 2 to 6 of the Health and Safety at Work etc. Act 1974. Commitment to continuous improvement involves the constant development of policies, approaches to implementation and techniques of risk control. Organisations which achieve high standards of health and safety assess their health and safety performance by internal reference to key performance indicators and by external comparison with the performance of business competitors. They often also record and account for their performance in their annual reports.

 1(d) Approved code of practice
   Management of health and safety at work
 1(e) HSE guidance notes
   Successful health and safety management
 3(a) Tables and figures
   Key elements of successful health and safety management.

Technique for human error rate probability (THERP)

Many accidents are associated with human error. THERP is a technique for predicting the potential for human error in a work activity. It evaluates quantitatively the contribution of the human error element in the development of an untoward incident.

The technique uses human behaviour as the basic unit of evaluation. It involves the concept of a basic error rate that is relatively consistent between tasks requiring similar human performance elements in different situations. Basic error rates are assessed in terms of contributions to specific systems failures.

The methodology of THERP entails:
  (a) selecting the system failure;
  (b) identifying all behaviour elements;
  (c) estimating the probability of human error; and
  (d) computing the probabilities as to which specific human error will produce the system failure.
Following classification of probable errors, specific corrective actions are introduced to reduce the likelihood of error.

The major weakness in the use of the THERP technique, however, is the lack of sufficient error rate data.

Total loss control

Total loss control is a management system developed in the 1960s by Frank Bird. It is defined as a programme designed to reduce or eliminate all accidents which downgrade the system and which result in wastage of an organisation's assets. An organisation's assets are:

Manpower – Materials – Machinery – Manufactured goods – Money (The five 'Ms')

Within the total loss control concept a number of definitions are important.

Incident
An undesired event that could, or does, result in loss

or

An undesired event that could, or does, downgrade the efficiency of the business operation.

Accident
An undesired event that results in physical harm or damage to property. It is usually the result of contact with a source of energy (i.e. kinetic, electrical, thermal, ionising, non-ionising radiation, etc.) above the threshold limit of the body or structure.

Loss control
An intentional management action directed at the prevention, reduction or elimination of the pure (non-speculative) risks of business.

Total loss control
The application of professional management techniques and skills through those programme activities (directed at risk avoidance, loss prevention and loss reduction) specifically intended to minimise loss resulting from the pure (non-speculative) risks of business.

Total loss control programmes
Total loss control is commonly run as a programme over a period of, for example, five years. The various stages are outlined below.

Injury prevention
This stage is concerned with the humanitarian and, to some extent, legal aspects of employee safety and employees' compensation costs. It normally incorporates a range of features, such as machinery safety, joint consultation, safety training, cleaning and housekeeping, safety rules, etc.

Damage control

This part of the programme covers the control of accidents which cause damage to property and plant and which might, conceivably, cause injury. Essential elements of this stage are damage reporting, recording and costing.

Total accident control

This stage of the programme is directed at the prevention of *all* accidents resulting in personal injury and/or property damage. Three important aspects of this stage are spot checking systems, reporting by control centres and health and safety audits.

Business interruption

This entails the incorporation in the programme of controls over all situations and influences which downgrade the system and result in interruption of the business activities, e.g. fire prevention, security procedures, product liability, pollution prevention. Business interruption results in lost money, e.g. operating expenses, lost time, reduced production and lost sales.

Total loss control
This is the control of all insured and uninsured costs arising from any incidents which downgrade the system.

The various stages of total loss control are shown below.

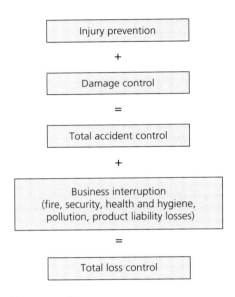

Stages of total loss control.

2(b)
Hazard checklists

This section incorporates a number of checklists which may be used in:
- safety monitoring activities;
- investigation of incidents; and
- as part of the risk assessment process.

Construction activities

1. Health and welfare	Yes	No
1.1 Are first aid boxes provided, maintained and readily available?	☐	☐
1.2 Are arrangements for calling an ambulance in place?	☐	☐
1.3 Is a responsible person appointed to take charge of situations where a first aider is not available?	☐	☐
1.4 Is a stretcher readily available?	☐	☐
1.5 Has a foul weather shelter been provided?	☐	☐
1.6 Is a mess room provided?	☐	☐
1.7 Are adequate and suitable clothing storage and changing facilities provided?	☐	☐
1.8 Are facilities provided for heating food?	☐	☐
1.9 Are sanitation arrangements (urinals, water closets) adequate?	☐	☐
1.10 Are washing and showering facilities adequate?	☐	☐
1.11 Are adequate facilities for rest provided?	☐	☐

(Continued)

Construction activities (*Continued*)

		Yes	No
1.12	Is a suitable supply of drinking water provided and suitably marked?	☐	☐
1.13	Is there a formal emergency procedure in place?	☐	☐

2. Environmental factors

		Yes	No
2.1	Is safe access and egress provided to the site?	☐	☐
2.2	Is a good standard of housekeeping and cleanliness maintained?	☐	☐
2.3	Is site lighting, including emergency lighting, provided and maintained?	☐	☐
2.4	Is there adequate segregation from non-construction activities?	☐	☐
2.5	Are adequate measures taken to control emissions of dust, fumes, etc?	☐	☐
2.6	Is effective ventilation provided in working areas?	☐	☐
2.7	Are ventilation arrangements adequate and effective?	☐	☐
2.8	Is there a system for ensuring adequate temperature control in indoor workplaces?	☐	☐
2.9	Are suitable and sufficient lighting arrangements provided and maintained?	☐	☐
2.10	Are measures taken to prevent adverse lighting conditions?	☐	☐
2.11	Is secondary lighting provided and maintained?	☐	☐
2.12	Are waste storage and disposal arrangements adequate?	☐	☐
2.13	Are perimeter signs installed and maintained?	☐	☐

3. Fire prevention and protection

		Yes	No
3.1	Is there adequate access for fire brigade appliances?	☐	☐

(*Continued*)

Construction activities (*Continued*)

		Yes	No
3.2	Is there a formal system for summoning fire brigade?	☐	☐
3.3	Are huts sited in safe positions?	☐	☐
3.4	Is there adequate space between huts?	☐	☐
3.5	Are huts of fireproof construction?	☐	☐
3.6	Are prohibited areas identified and suitable notices displayed?	☐	☐
3.7	Have specific fire risks been identified and assessed?	☐	☐
3.8	Are measures taken to control the use of flame-producing plant and equipment?	☐	☐
3.9	Are heaters in huts installed safely and heating arrangements maintained?	☐	☐
3.10	Are fire appliances installed at suitable locations?	☐	☐
3.11	Are fire detectors and alarms installed and maintained?	☐	☐
3.12	Is there specific provision for the storage of flammable substances?	☐	☐
3.13	Are vehicle parking arrangements satisfactory from a fire protection viewpoint?	☐	☐
4. Storage of materials			
4.1	Are materials sited in safe areas?	☐	☐
4.2	Are materials stacked safely?	☐	☐
4.3	Are storage huts provided and maintained?	☐	☐
4.4	Are flammable materials adequately separated?	☐	☐
4.5	Are compressed gases stored safely in a secure compound?	☐	☐
4.6	Are compressed gases adequately segregated?	☐	☐

(*Continued*)

Construction activities (*Continued*)

	Yes	No
4.7 Are provisions for the storage of hazardous substances adequate?	☐	☐
4.8 Are provisions for the storage of explosives adequate?	☐	☐
5. Plant, machinery and hand tools		
5.1 Are: • lifting equipment and appliances • woodworking machinery • electrical equipment • abrasive wheels • welding equipment • hand tools maintained in an efficient state, in efficient working order and in good repair?	☐ ☐ ☐ ☐ ☐ ☐ ☐	☐ ☐ ☐ ☐ ☐ ☐ ☐
5.2 Are formal procedures for the maintenance, examination and testing of plant and machinery in place?	☐	☐
5.3 Are guarding and fencing arrangements adequate and suitably maintained?	☐	☐
5.4 Are measures taken to ensure that plant and machinery are of adequate construction, strength and suitability?	☐	☐
6. Access equipment and working places		
6.1 Is all scaffolding correctly erected?	☐	☐
6.2 Are all ladders in good condition, suitably placed and secured at their upper resting place?	☐	☐
6.3 Are all trenches and excavations maintained in a safe condition?	☐	☐
6.4 Are measures taken to protect employees from risks from overhead cables?	☐	☐
6.5 Are all working platforms suitably protected by means of a top rail, intermediate rail and toe board?	☐	☐

(*Continued*)

Construction activities (*Continued*)

		Yes	No
6.6	Is movable access equipment adequately maintained and moved under safe conditions?	☐	☐
6.7	Are adequate measures taken to prevent falls arising from work on fragile materials?	☐	☐
6.8	Are measures taken to prevent risks arising from falling objects?	☐	☐
6.9	Are adequate means provided for arresting falls?	☐	☐
6.10	Is personal suspension equipment suitably maintained?	☐	☐
6.11	Are any unstable structures identified and access to same prevented?	☐	☐
6.12	Have competent persons been appointed with respect to the inspection of scaffolds, excavations and personal suspension equipment?	☐	☐

7. Cofferdams and caissons

		Yes	No
7.1	Are cofferdams and caissons suitably designed and constructed?	☐	☐
7.2	Are the materials suitable in each case?	☐	☐
7.3	Is the strength and capacity of materials adequate?	☐	☐
7.4	Are adequate maintenance arrangements in place?	☐	☐
7.5	Has a competent person been appointed to undertake inspections of cofferdams and caissons?	☐	☐

8. Prevention of drowning

		Yes	No
8.1	Have the hazards associated with work over water been identified?	☐	☐
8.2	Have the risks arising from drowning been assessed?	☐	☐

(*Continued*)

Construction activities (*Continued*)

		Yes	No
8.3	Is rescue equipment adequate and readily available for use in an emergency?	☐	☐
8.4	Where work entails transport over water, have the hazards been identified?	☐	☐
8.5	Are vessels suitable and adequately maintained?	☐	☐
8.6	Have flooding risks been identified and assessed?	☐	☐
9. Radiography and radioactive materials			
9.1	Have classified workers been identified and trained?	☐	☐
9.2	Are medical supervision arrangements adequate?	☐	☐
9.3	Are personal dose monitoring arrangements adequate?	☐	☐
9.4	Are sealed sources of radiation suitably controlled?	☐	☐
9.5	Are personal dose records maintained?	☐	☐
9.6	Are personal dose meters/film badges provided and used?	☐	☐
9.7	Has a competent person been appointed to oversee preventive and protective measures?	☐	☐
10. Site transport			
10.1	Are cautionary signs and notices adequate and displayed prominently?	☐	☐
10.2	Are directional signs adequate and displayed prominently?	☐	☐
10.3	Do site layout arrangements ensure safe movement of people and vehicles?	☐	☐
10.4	Are authorised drivers appointed in the case of certain site vehicles?	☐	☐
10.5	Are lift trucks used safely?	☐	☐
10.6	Is mobile access equipment used safely?	☐	☐

(*Continued*)

Construction activities (*Continued*)

		Yes	No
10.7	Are potentially dangerous vehicles excluded from the site?	☐	☐
10.8	Are disciplinary measures in place in the event of unsafe driving?	☐	☐
10.9	Is access to and egress from site safe and adequately controlled?	☐	☐
10.10	Are separate parking areas provided for site employees?	☐	☐
10.11	Are pedestrian and vehicular traffic routes adequately segregated?	☐	☐
10.12	Are all traffic routes suitable for their purpose and adequately maintained?	☐	☐
10.13	Are traffic routes kept clear at all times?	☐	☐
10.14	Are the arrangements for vehicle towing safe?	☐	☐
10.15	Are passenger-carrying vehicles safe and adequately maintained?	☐	☐
10.16	Are measures taken to ensure the safe loading of vehicles?	☐	☐
10.17	Are measures provided to prevent overrunning?	☐	☐
10.18	Are emergency routes and exits clearly identified and maintained?	☐	☐

11. Personal protective equipment

		Yes	No
11.1	Are safety helmets provided and worn at all times?	☐	☐
11.2	Are the following items of personal protective equipment provided and maintained:		
	• eye/face protection?	☐	☐
	• hand and arm protection?	☐	☐
	• respiratory protection?	☐	☐
	• leg and foot protection?	☐	☐
	• foul weather clothing?	☐	☐
	• ear protection?	☐	☐
	• body protection?	☐	☐

(*Continued*)

Construction activities (*Continued*)

		Yes	No
11.3	Is personal suspension equipment provided and maintained?	☐	☐
12. Demolition			
12.1	Has a pre-demolition survey been undertaken?	☐	☐
12.2	Have method statements been prepared and implemented correctly?	☐	☐
12.3	Have all services been isolated?	☐	☐
12.4	Has the demolition area been segregated from the rest of the site by physical barriers?	☐	☐
12.5	Have fans and catching platforms been installed where necessary?	☐	☐
12.6	Have temporary props been installed where necessary?	☐	☐
12.7	Is there adequate control over pulling arrangements, use of demolition ball and pushing arrangements?	☐	☐
12.8	Are formal precautions in place for the use of explosives?	☐	☐
12.9	Have suitable measures been installed to provide protection against falling, i.e. harnesses, safety nets?	☐	☐
12.10	Are measures for the protection against falling items adequate?	☐	☐
12.11	Is access to dangerous areas adequately controlled?	☐	☐
12.12	Are the precautions for work over open joisting adequate?	☐	☐
12.13	Has the glass in windows, doors and partitions been removed prior to demolition?	☐	☐
12.14	Are scaffolds provided and used for manual demolition?	☐	☐

(*Continued*)

Construction activities (*Continued*)

	Yes	No
12.15 Have express measures been taken to prevent premature collapse?	☐	☐
12.16 Are drivers/operators and banksmen adequately trained?	☐	☐
12.17 Is adequate lighting provided?	☐	☐
12.18 Has appropriate personal protective equipment been provided for all persons?	☐	☐
12.19 Have suitable measures been taken to detect the presence of asbestos?	☐	☐
12.20 Has a competent person been appointed to oversee demolition operations?	☐	☐

13. Personnel

	Yes	No
13.1 Have measures been taken to ensure the provision of:		
• safety awareness training?	☐	☐
• first aid training?	☐	☐
• competent person training?	☐	☐
13.2 Has the need for health surveillance of certain employees been assessed and health surveillance provided where necessary?	☐	☐

14. Inspection and reports

	Yes	No
14.1 Are procedures in operation to ensure the frequent inspection of:		
• guard rails etc.?	☐	☐
• working platforms?	☐	☐
• excavations?	☐	☐
• personal suspension equipment?	☐	☐
• means for arresting falls?	☐	☐
• ladders?	☐	☐
• cofferdams and caissons?	☐	☐
• welfare facilities?	☐	☐
14.2 Are reporting arrangements in place in respect of the above inspections?	☐	☐

Display screen equipment workstations

1. Equipment	Yes	No
Display screen		
1.1 Are the characters on the screen well-defined and clearly formed, of adequate size and with adequate spacing between the characters and lines?	☐	☐
1.2 Is the image on the screen stable, with no flickering or other forms of instability?	☐	☐
1.3 Is brightness and the contrast between the characters and the background easily adjustable by the user or operator, and easily adjustable to ambient conditions?	☐	☐
1.4 Does the screen swivel and tilt easily and freely to suit the needs of the user/operator?	☐	☐
1.5 Is it possible to use a separate base for the screen or an adjustable table?	☐	☐
1.6 Is the screen free of reflective glare and reflections liable to cause discomfort to the user/operator?	☐	☐
Keyboard		
1.7 Is the keyboard tiltable and separate from the screen so as to allow the user/operator to find a comfortable working position avoiding fatigue in the arms or hands?	☐	☐
1.8 Is the space in front of the keyboard sufficient to provide support for the hands and arms of the user/operator?	☐	☐
1.9 Does the keyboard have a matt surface to avoid reflective glare?	☐	☐
1.10 Are the arrangement of the keyboard and the characteristics of the keys such as to facilitate the use of the keyboard?	☐	☐

(*Continued*)

Display screen equipment workstations (*Continued*)

		Yes	No
1.11	Are the symbols on the keys adequately contrasted and legible from the design working position?	☐	☐
Work desk or work surface			
1.12	Does the work desk or work surface have a sufficiently large, low reflectance surface and allow a flexible arrangement of the screen, keyboard, documents and related equipment?	☐	☐
1.13	Is the document holder stable and adjustable and so positioned as to minimise the need for uncomfortable head and eye movements?	☐	☐
1.14	Is there adequate space for users/operators to find a comfortable position?	☐	☐
Work chair			
1.15	Is the work chair stable and does it allow the user/operator easy freedom of movement and a comfortable position?	☐	☐
1.16	Is the seat adjustable in height?	☐	☐
1.17	Is the seat back adjustable in both height and tilt?	☐	☐
1.18	Is a footrest made available to any user/operator who wishes one?	☐	☐
2. Environmental factors			
Space requirements			
2.1	Is sufficient space provided for the user/operator to change position and vary movements?	☐	☐
Lighting			
2.2	Is there adequate ambient lighting at the workstation?	☐	☐
2.3	Is there adequate contrast between the screen and the background environment?	☐	☐

(*Continued*)

Display screen equipment workstations (*Continued*)

		Yes	No
2.4	Are disturbing glare and reflections on the screen or other equipment prevented by co-ordinating workplace and workstation layout with the positioning and technical characteristics of the artificial light sources?	☐	☐

Reflections and glare

		Yes	No
2.5	Is the workstation so designed that sources of light, such as windows and other openings, transparent or translucid walls, and brightly coloured fixtures or walls cause no direct glare and no distracting reflections on the screen?	☐	☐
2.6	Are windows fitted with a suitable system of adjustable covering to attenuate the daylight that falls on the workstation?	☐	☐

Noise

2.7	Is noise emitted by equipment belonging to the workstation taken into account when the workstation is being equipped, with a view in particular to ensuring that attention is not distracted and speech is not disturbed?	☐	☐

Heat

2.8	Are measures taken to ensure that excess heat from equipment does not cause discomfort to users or operators?	☐	☐

Radiation

2.9	Are measures taken to ensure that all radiation is reduced to negligible levels from the point of view of the protection of the user/operator's health and safety?	☐	☐

Humidity

2.10	Is an adequate level of humidity maintained at the workstation?	☐	☐

(*Continued*)

Display screen equipment workstations (*Continued*)

3. Interface between computer and user/operator	Yes	No
Software and systems		
3.1 Is software: • suitable for the task? • easy to use and, where appropriate adaptable to the level of knowledge or experience of the operator or user?	☐ ☐	☐ ☐
3.2 If a quantitative or qualitative checking facility is used, are users or operators advised of the use of this facility?	☐	☐
3.3 Do systems provide feedback to users or operators on the performance of those systems?	☐	☐
3.4 Do systems display information in a format and at a pace which are adapted to users or operators?	☐	☐
3.5 Are the principles of software ergonomics applied, in particular to human data processing?	☐	☐

Electrical equipment

		Yes	No
1.1	Is there adequate protection against direct and indirect contact?	☐	☐
1.2	Is the equipment earthed?	☐	☐
1.3	Is a fuse installed and of the correct rating?	☐	☐
1.4	Does the system incorporate a circuit breaker?	☐	☐
1.5	Does the system incorporate an earth leakage circuit breaker (residual current device)?	☐	☐
1.6	Does the system operate at reduced voltage?	☐	☐
1.7	Does use of the equipment entail the need for a safe system of work (permit-to-work system) in terms of:		
	• isolation procedure?	☐	☐
	• checking voltages?	☐	☐
	• supervision and control arrangements?	☐	☐
	• physical precautions, e.g. barriers?	☐	☐
1.8	Are conductors insulated or safe by position?	☐	☐
1.9	Are joints and connections mechanically and electrically suitable?	☐	☐
1.10	Is there means for protection against excess current?	☐	☐
1.11	Is there means for cutting off electrical supply and isolation of any circuit?	☐	☐
1.12	Are precautions taken with respect to equipment made dead?	☐	☐
1.13	Is there adequate working space, means of access and lighting?	☐	☐
1.14	Are arrangements for portable appliance testing adequate?	☐	☐
1.15	Is a portable appliance register available and maintained?	☐	☐

(Continued)

Electrical equipment (*Continued*)

		Yes	No
1.16	Have competent person(s) been appointed?	☐	☐
1.17	Are all suspect portable appliances removed from service immediately?	☐	☐
1.18	Is equipment visually inspected on a frequent basis?	☐	☐
1.19	Are measures taken to ensure the safe use of portable appliances?	☐	☐
1.20	Are appropriate precautions taken where work may entail adverse or hazardous environments?	☐	☐

Fire safety

1. Electrical appliances	Yes	No
1.1 Are appliances inspected, tested and labelled with date of last inspection?	☐	☐
1.2 Are heavy load appliances, including space heaters, connected to a permanent outlet?	☐	☐
1.3 Are flexes visible at all times?	☐	☐
1.4 Are circuit breakers of appropriate capacity installed?	☐	☐
1.5 Is the use of multi-point adaptors prohibited?	☐	☐
1.6 Are measures taken to prevent unsafe wiring and connections?	☐	☐
2. Fire doors and partitions		
2.1 Are doors and partitions in sound condition?	☐	☐
2.2 Are self-closures fitted to doors?	☐	☐
2.3 Are fire doors kept free from obstruction at all times?	☐	☐
2.4 Are flame-retardant partitions installed at critical points?	☐	☐
2.5 Are smoke detectors and sprinkler heads clear of obstructions?	☐	☐
3. Decorations and furnishings		
3.1 Are curtains and other fabrics flame-proofed?	☐	☐
3.2 Is there a prohibition on the use of flammable decorations?	☐	☐
4. Open flame devices and appliances		
4.1 Is there a prohibition on the use of candles, gas lamps or other open flame devices?	☐	☐
4.2 Are 'hot work' activities well-controlled?	☐	☐

(Continued)

Fire safety (*Continued*)

	Yes	No
5. General storage		
5.1 Are storage areas maintained in a clean and orderly condition?	☐	☐
5.2 Is flammable refuse stored in closed metal bins?	☐	☐
5.3 Are electric panels, fire appliances, fire exits, fire detection devices and sprinkler heads kept free from obstruction?	☐	☐
5.4 Is there a prohibition on storage of articles on stairways and in unauthorised areas?	☐	☐
6. Flammable and combustible substances		
6.1 Is there adequate ventilation to prevent accumulation of vapours?	☐	☐
6.2 Have all sources of ignition been eliminated?	☐	☐
6.3 Are small containers of flammable liquids stored in metal cabinets?	☐	☐
6.4 Are large containers e.g. 50 gallon drums, of flammable liquids stored in purpose-built store?	☐	☐
7. Smoking		
7.1 Are there restrictions on smoking, including designated 'No smoking' areas?	☐	☐
7.2 Is there evidence of smoking in unauthorised areas?	☐	☐
8. Means of escape and exits		
8.1 Is the means of escape clearly indicated?	☐	☐
8.2 Are fire exit doors maintained and capable of opening with ease?	☐	☐
8.3 Are fire exits kept clear at all times and suitably marked?	☐	☐
8.4 Is there adequate provision for disabled persons and wheelchair users?	☐	☐

(*Continued*)

Fire safety (*Continued*)

9. Housekeeping and cleaning	Yes	No
9.1 Is a high level of housekeeping maintained?	☐	☐
9.2 Is there adequate storage and segregation of flammable refuse?	☐	☐
9.3 Is the layout of working areas safe?	☐	☐
9.4 Is cleaning and removal of refuse regularly carried out?	☐	☐
10. Fire appliances		

10.1 Are fire appliances:
   • clearly indicated?
   • wall-mounted?
   • maintained?
	
	☐
	☐
	☐
		
		☐
		☐
		☐

10.2 Have designated employees been trained in the correct use of appliances?	☐	☐
11. Fire instructions		
11.1 Are fire instructions clearly displayed?	☐	☐
11.2 Are fire instructions drawn to the attention of employees regularly?	☐	☐
12. Fire alarm and detection devices		
12.1 Are alarms and detection devices clearly indicated?	☐	☐
12.2 Are employees trained in responding to the alarm?	☐	☐
12.3 Are alarms and detection devices maintained?	☐	☐
12.4 Are detection devices wired in?	☐	☐
13. Sprinkler systems		
13.1 Is the system inspected and maintained on a regular basis?	☐	☐
13.2 Are sprinkler heads kept unobstructed?	☐	☐

Flammable substances

1. Flammable liquids	Yes	No
1.1 Are there separate storage arrangements for flammable liquids?	☐	☐
1.2 Are only the smallest quantities stored in work area?	☐	☐
1.3 Are small quantities of flammable liquids transported in closed transfer containers?	☐	☐
1.4 Are transfer containers correctly labelled?	☐	☐
1.5 Is the system for dispensing from bulk containers safe?	☐	☐
1.6 Are fire appliances available during use and dispensing operations?	☐	☐
1.7 Is there adequate ventilation in storage rooms?	☐	☐
1.8 Is a ban on smoking and naked lights actively enforced?	☐	☐
2. Liquefied and compressed gases		
2.1 Are cylinders stored and transported in the upright position?	☐	☐
2.2 Are cylinders stored in open well-ventilated areas out of direct sunlight?	☐	☐
2.3 Are cylinders secured with wall chains or stored in racks?	☐	☐
2.4 Are storage areas and rooms suitably marked?	☐	☐
2.5 Are internal store rooms:		
• in safe position or of fire-resisting structure?	☐	☐
• adequately ventilated?	☐	☐
• used solely for storage of LPG and/or acetylene cylinders?	☐	☐
• provided with safe means of escape?	☐	☐
2.6 Are there adequate precautions against spillages and leakages?	☐	☐

(Continued)

Flammable substances (*Continued*)

		Yes	No
2.7	Are employees supervised to ensure no handling of cylinders by their valves and no dropping or rolling of cylinders?	☐	☐
2.8	Are there measures to ensure all cylinder valves are turned off when not in use?	☐	☐
2.9	Are oxygen cylinders stored separately?	☐	☐
2.10	Are measures taken to prevent build up of solid waste residues on surfaces?	☐	☐
2.11	Is fire-fighting equipment readily available?	☐	☐
2.12	Are procedures in operation for the reporting of defects in plant, equipment and appliances?	☐	☐

Floors and traffic routes

		Yes	No
1.1	Are all floors and traffic routes suitable for the purpose for which they are used?	☐	☐
1.2	Are all floors and traffic routes of sound construction, adequate strength and stability, taking into account the loads placed on them and the traffic passing over them?	☐	☐
1.3	Are measures taken to ensure that the floor or surface of a traffic route has no dangerous hole or slope, and is even and slip-proof?	☐	☐
1.4	Are holes and defects in surfaces to floors and traffic routes dealt with expeditiously?	☐	☐
1.5	Are temporary holes in floors and traffic routes suitably protected against the risk of tripping or falling?	☐	☐
1.6	Is floor drainage provided where floors are liable to be wet on a regular basis?	☐	☐
1.7	Are the surfaces of floors and traffic routes which are likely to get wet regularly, or be subject to spillages, slip resistant?	☐	☐
1.8	Are floors and traffic routes kept free from obstructions and from articles and substances which could contribute to slips and falls?	☐	☐
1.9	Are floors around machinery slip-resistant and kept free from slippery substances and loose materials?	☐	☐
1.10	Are measures taken to keep floors and traffic routes free of obstruction?	☐	☐
1.11	Are housekeeping procedures to cover spillages adequate?	☐	☐
1.12	Are measures taken to prevent slipping in the event of snow and ice?	☐	☐
1.13	Are handrails and, where appropriate, guard rails, provided on staircases?	☐	☐
1.14	Where floors are liable to become slippery are employees provided with slip-resistant footwear?	☐	☐

Hazardous substances

1. Information and identification	Yes	No
1.1 Is an up-to-date list of all hazardous substances held on site readily available?	☐	☐
1.2 Are safety data sheets available for all hazardous substances held on site?	☐	☐
1.3 Are all packages and containers correctly labelled?	☐	☐
1.4 Are all transfer containers suitable for the purpose and correctly labelled?	☐	☐
2. Storage		
2.1 Are all stores and external storage areas safe in respect of construction, layout, security and control?	☐	☐
2.2 Are hazardous substances correctly segregated?	☐	☐
2.3 Are cleaning and housekeeping levels adequate?	☐	☐
2.4 Are all issues of hazardous substances to employees adequately controlled?	☐	☐
3. Protection		
3.1 Are appropriate warning signs displayed in areas of storage and use?	☐	☐
3.2 Is suitable personal protective equipment available, maintained in safe condition and used by employees?	☐	☐
3.3 Are emergency eye wash facilities and showers available, suitably located and maintained?	☐	☐
3.4 Are the above facilities provided with frost protection?	☐	☐
3.5 Are adequate and suitable first aid facilities available, suitably located and maintained?	☐	☐
3.6 Are adequate and appropriate fire appliances available, suitably located, accessible and maintained?	☐	☐

(*Continued*)

Hazardous substances (*Continued*)

	Yes	No
3.7 Are supplies of appropriate neutralising compounds readily available in the event of spillage?	☐	☐
4. Procedures		
4.1 Are written safe handling procedures prepared and available for all hazardous substances?	☐	☐
4.2 Is there a specific procedure for dealing with spillages of all types of substance?	☐	☐
4.3 Is there a routine inspection procedure covering: • personal protective equipment? • emergency showers and eye washes? • first aid boxes? • fire appliances? • chemical dosing to plant? • use of neutralising compounds?	☐ ☐ ☐ ☐ ☐ ☐	☐ ☐ ☐ ☐ ☐ ☐
5. Training		
5.1 Are employees trained in: • safe handling procedures? • use of fire appliances? • dealing with spillages? • the use and care of personal protective equipment?	☐ ☐ ☐ ☐	☐ ☐ ☐ ☐
5.2 Are first aiders trained to deal with injuries associated with hazardous substances?	☐	☐
5.3 Are training records maintained?	☐	☐

Maintenance work

	Yes	No
1.1 Is a safe system of work, including operation of permit-to-work system in high risk situations, established, documented and used?	☐	☐
1.2 Are competent persons appointed for high risk operations?	☐	☐
1.3 Are method statements prepared and used, particularly for contractors involved in maintenance work?	☐	☐
1.4 Are 'Rules for the safe conduct of project work' actively enforced?	☐	☐
1.5 Are controlled areas specifically designated?	☐	☐
1.6 Is there adequate control over access to the working area?	☐	☐
1.7 Have all operators been provided with information, instruction and training?	☐	☐
1.8 Are all signs, marking and labelling adequate?	☐	☐
1.9 Is personal protective equipment provided, suitable for the circumstances and used?	☐	☐

Manual handling operations

The following check list, which forms Schedule 1 to the Manual Handling Operations Regulations 1992, should be used.

Factors to which the employer must have regard and questions he must consider when making an assessment of manual handling operations.

1. The tasks

	Yes	No
Do they involve:		
• holding or manipulating loads at distance from trunk?	☐	☐
• unsatisfactory bodily movement or posture, especially		
○ twisting the trunk?	☐	☐
○ stooping?	☐	☐
○ reaching upwards?	☐	☐
• excessive movement of loads, especially:		
○ excessive lifting or lowering distances?	☐	☐
○ excessive carrying distances?	☐	☐
• excessive pushing or pulling of loads?	☐	☐
• risk of sudden movement of loads?	☐	☐
• frequent or prolonged physical effort?	☐	☐
• insufficient rest or recovery periods?	☐	☐
• a rate of work imposed by a process?	☐	☐

2. The loads

Are they:	Yes	No
• heavy?	☐	☐
• bulky or unwieldy?	☐	☐
• difficult to grasp?	☐	☐
• unstable, or with contents likely to shift?	☐	☐
• sharp, hot or otherwise potentially damaging?	☐	☐

3. The working environment

Are there:	Yes	No
• space constraints preventing good posture?	☐	☐
• uneven, slippery or unstable floors?	☐	☐
• variations in level of floors or work surfaces?	☐	☐

(Continued)

Manual handling operations (*Continued*)

	Yes	No
• extremes of temperature or humidity?	☐	☐
• conditions causing ventilation problems or gusts of wind?	☐	☐
• poor lighting conditions?	☐	☐

4. Individual capability

Does the job:
- require unusual strength, height, etc.? ☐ ☐
- create a hazard to those who might reasonably be considered to be pregnant or to have a health problem? ☐ ☐
- require special information or training for its safe performance? ☐ ☐

5. Other factors

Is movement or posture hindered by personal protective equipment or by clothing? ☐ ☐

Mobile mechanical handling equipment (lift trucks, etc.)

		Yes	No
1.1	Is the equipment used only by authorised and trained operators?	☐	☐
1.2	Where left unattended, are the forks of the truck lowered and controls immobilised?	☐	☐
1.3	Are drivers trained and supervised to ensure the maximum rated capacity of the truck is not exceeded?	☐	☐
1.4	Are trucks adapted or equipped to reduce the risk to the operator from overturning?	☐	☐
1.5	Does each truck incorporate a roll-over protective structure which ensures it does no more than fall on its side?	☐	☐
1.6	Do trucks incorporate devices for improving the operators' vision?	☐	☐
1.7	Are measures taken to ensure passengers are not carried unless in a specific cage or purpose designed working platform?	☐	☐
1.8	When used on public highway, are trucks fitted with appropriate brakes, lights and steering?	☐	☐
1.9	Are keys kept securely and issued only to authorised operators?	☐	☐
1.10	Are trucks constructed or adapted as to be suitable for the purpose of use?	☐	☐
1.11	Are trucks used only for operations which, and under conditions for which, they are suitable?	☐	☐
1.12	Are all trucks maintained in an efficient state, in efficient working order and in good repair?	☐	☐
1.13	Are trucks subject to a formal maintenance programme based on the manufacturers'	☐	☐

(Continued)

Mobile mechanical handling equipment (lift trucks, etc.) (*Continued*)

	Yes	No
servicing recommendations for inspection, maintenance and servicing?		
1.14 Are all trucks subject to a weekly maintenance check including an examination of steering gear, lifting gear, brakes, lighting, battery, mast, forks, attachments and any chains or ropes used in the lifting mechanism?	☐	☐
1.15 Are lifting chains examined on an annual basis and certificated?	☐	☐
1.16 Are trucks subject to six-monthly and annual examination by a trained fitter or representative of the manufacturer?	☐	☐
1.17 Is a maintenance log kept up to date for each truck?	☐	☐

Noise

	Yes	No
1.1 Is exposure to workplace noise assessed on a regular basis by means of:		
(a) observation of specific working practices?	☐	☐
(b) reference to relevant information on the probable levels of noise corresponding to any equipment used in the particular working conditions?	☐	☐
(c) if necessary, measurement of the level of noise to which employees are likely to be exposed?	☐	☐
1.2 As a result of assessments, are any employees likely to be exposed to noise at or above a lower exposure action value, an upper exposure action value, or an exposure limit value?	☐	☐
1.3 Do risk assessments include consideration of:		
(a) the level, type and duration of exposure, including any exposure to peak sound pressure?	☐	☐
(b) the effects of exposure to noise on employees or groups of employees whose health is at particular risk from such exposure?	☐	☐
(c) any effects on the health and safety of employees resulting from the interaction between noise and the use of ototoxic substances at work, or between noise and vibration?	☐	☐
(d) any indirect effects on the health and safety of employees resulting from the interaction between noise and audible warning signals or other sounds that need to be audible in order to reduce risk at work?	☐	☐
(e) any information provided by the manufacturers of work equipment?	☐	☐
(f) the availability of alternative equipment designed to reduce the emission of noise?	☐	☐
(g) any extension of exposure to noise at the workplace, including exposure in rest facilities supervised by the employer?	☐	☐

(*Continued*)

Noise (*Continued*)

		Yes	No
(h)	appropriate information obtained following health surveillance, including, where possible, published information?	☐	☐
(i)	the availability of personal hearing protectors with adequate attenuation characteristics?	☐	☐
1.4	Are records of assessments maintained and reviewed regularly?	☐	☐
1.5	Is the risk of hearing damage either eliminated at source or reduced to as low a level as is reasonably practicable?	☐	☐
1.6	If an employee is likely to be exposed to noise at or above an upper exposure action value, is exposure reduced to as a low a level as is reasonably practicable by establishing and implementing a programme or organisational and technical measures, excluding the provision of personal hearing protectors, which is appropriate to the activity?	☐	☐
1.7	Do the actions taken by the employer include consideration of:		
(a)	other working methods which reduce exposure to noise?	☐	☐
(b)	choice of appropriate work equipment emitting the least possible noise, taking account of the work to be done?	☐	☐
(c)	the design and layout of workplaces, work stations and rest facilities?	☐	☐
(d)	suitable and sufficient information and training for employees, such that work equipment may be used correctly, in order to minimise their exposure to noise?	☐	☐
(e)	reduction of noise by technical means?	☐	☐
(f)	appropriate maintenance programmes for work equipment, the workplace and workplace systems?	☐	☐
(g)	limitation of the duration and intensity of exposure to noise?	☐	☐

(*Continued*)

Noise (*Continued*)

		Yes	No
	(h) appropriate work schedules with adequate rest periods?	☐	☐
1.8	Is personal hearing protection provided for employees where they are likely to be exposed to noise at or above a lower exposure action value?	☐	☐
1.9	Are Hearing Protection Zones identified and demarcated?	☐	☐
1.10	Is entry to Hearing Protection Zones controlled?	☐	☐
1.11	Are noise control measures: (a) fully and properly used? (b) maintained in an efficient state, in efficient working order and in good repair?	☐ ☐	☐ ☐
1.12	Do employees: (a) make full and proper use of personal hearing protectors and any control measures provided? (b) report defects in personal hearing protectors or other control measures?	☐ ☐	☐ ☐
1.13	Where risk assessment identifies a risk to health to employees, are such employees placed under suitable health surveillance, including testing of their hearing?	☐	☐
1.14	Are health records maintained for those employees undergoing health surveillance?	☐	☐
1.15	Where employees are found to have identifiable hearing damage, does the employer ensure they are examined by a doctor and/or appropriate specialist?	☐	☐
1.16	Are employees provided with suitable information, instruction and training where they may be exposed to noise at or above a lower action value?	☐	☐

Offices and commercial premises

1. Fire	Yes	No
1.1 Is there evidence of overloading, abuse and misuse of electrical systems?	☐	☐
1.2 Are flammable materials, such as floor polishes, spirit-based cleaning fluids and packing materials stored safely?	☐	☐
1.3 Is smoking prohibited?	☐	☐
1.4 Is there evidence of employees smoking in store rooms and other areas not generally occupied by employees?	☐	☐
1.5 Is the use of freestanding heating appliances, particularly radiant-type electric fires and portable gas-fired heaters, prohibited?	☐	☐
1.6 Is there adequate control over the storage of waste paper and packing materials?	☐	☐
1.7 Is the fire alarm tested on a weekly basis?	☐	☐
1.8 Is an emergency fire evacuation undertaken regularly?	☐	☐
1.9 Are competent persons appointed and trained to oversee evacuations of the premises in the event of fire?	☐	☐
1.10 Are key personnel trained in the correct use of fire appliances?	☐	☐
2. Structural items		
2.1 Are measures taken to prevent slipping and tripping hazards arising from:		
• slippery floor finishes, wet floors, spillages and defective floor surfaces?	☐	☐
• dangerous staircases?	☐	☐
• the use by employees of unsuitable footwear?	☐	☐
2.2 Do all swing doors incorporate a clear view panel?	☐	☐

(Continued)

Offices and commercial premises (*Continued*)

	Yes	No
3. Work equipment		
3.1 Is all equipment suitable for the purpose for which it is used?	☐	☐
3.2 Is all equipment maintained in an efficient state, in efficient working order and in good repair?	☐	☐
4. Passenger lifts		
4.1 Are all lifts regularly inspected and maintained?	☐	☐
4.2 Do all lift cars level off correctly at their landings?	☐	☐
4.3 Are all lift cars marked with the maximum safe working load?	☐	☐
4.4 Do all lift cars incorporate a notice indicating the maximum number of passengers who can be carried at one time?	☐	☐
4.5 Do lift cars incorporate an emergency alarm device in the event of breakdown between floors or failure of the doors to open?	☐	☐
5. Access equipment		
5.1 Are all forms of access equipment, e.g. ladders, step ladders and mobile ladders, inspected on a regular basis?	☐	☐
5.2 Are employees instructed in the safe use of access equipment?	☐	☐
5.3 Are ladders for internal use fitted with non-slip feet?	☐	☐
5.4 Are step ladders fitted with a handrail?	☐	☐
5.5 Are working platforms to step ladders fitted with a guard rail?	☐	☐
6. Hand tools		
6.1 Are measures taken to ensure that safety knives are provided where necessary?	☐	☐

(*Continued*)

Offices and commercial premises (*Continued*)

		Yes	No
7.	**Manual handling operations**		
7.1	Are employees exposed to risk of manual handling injuries?	☐	☐
7.2	Are mechanical handling aids, such as hand trucks and trolleys, provided to eliminate the need for manual handling?	☐	☐
7.3	Are employees instructed to use the mechanical handling aids wherever practicable?	☐	☐
7.4	Are specific precautions taken in the case of pregnant female employees, young persons and physically disabled employees with respect to manual handling operations?	☐	☐
7.5	Has a suitable and sufficient manual handling risk assessment been undertaken in cases where potentially dangerous manual handling cannot be avoided? (See *Manual handling operations* check list, p. 229)	☐	☐
8.	**Hazardous substances**		
8.1	Are hazardous substances, such as cleaning preparations, used?	☐	☐
8.2	Have employees and other persons, such as the employees of cleaning contractors, been informed, instructed and trained on the safe use of such substances?	☐	☐
8.3	Are persons using hazardous substances adequately supervised?	☐	☐
9.	**Electricity**		
9.1	Have the premises been re-wired within the last 10 years?	☐	☐
9.2	Are all electrical appliances examined and tested by a competent person on a regular basis and a record of such inspections and tests maintained?	☐	☐

(*Continued*)

Offices and commercial premises (*Continued*)

		Yes	No
9.3	Is there evidence of insufficient electrical points resulting in the use of multi-point adaptors and extension leads?	☐	☐
9.4	Is there evidence of overloading of electrical sockets?	☐	☐
9.5	Are flexes, leads and cables: • in good condition and undamaged? • properly connected to the appliance? • securely attached to the plug? • not allowed to trail across floors? • where laid across floors, contained in protective strips?	☐ ☐ ☐ ☐ ☐	☐ ☐ ☐ ☐ ☐
9.6	Are all plugs fitted with the correct fuse?	☐	☐
9.7	Where more than one appliance is supplied from a power point, is a fused multi-socket block used?	☐	☐
9.8	Is there a procedure for ensuring faulty electrical equipment is removed from service immediately?	☐	☐

10. Display screen equipment workstations

See *Display screen equipment workstations* check list, p. 214

11. Environmental factors

11.1	Do employees have sufficient space to undertake tasks safely?	☐	☐
11.2	Is a reasonable temperature maintained at all times during working hours?	☐	☐
11.3	Is a thermometer provided on each floor to enable employees to determine the temperature of the workplace?	☐	☐
11.4	Is effective and suitable provision made to ensure that every part of the workplace is ventilated by a sufficient quantity of fresh or purified air?	☐	☐

(*Continued*)

Offices and commercial premises (*Continued*)

	Yes	No
11.5 Is suitable and sufficient lighting provided throughout the premises, including external areas?	☐	☐
11.6 Are measures taken to control excessive noise?	☐	☐
12. Cleaning and housekeeping		
12.1 Is the workplace (including furniture, furnishings and fittings) kept sufficiently clean?	☐	☐
12.2 Is there a formally written cleaning schedule for the premises?	☐	☐
12.3 Are cleaning activities monitored on a regular basis to ensure compliance with the cleaning schedule?	☐	☐
13. Health risks		
13.1 Are employees exposed to the risk of:		
• work-related upper limb disorders?	☐	☐
• visual fatigue?	☐	☐
• postural fatigue?	☐	☐
• stress-related ill-health?	☐	☐
14. Violence, bullying and harassment		
14.1 Is there a corporate policy dealing with violence, bullying and harassment at work?	☐	☐
14.2 Is there evidence that certain managers and supervisors may subject employees to mental and physical violence, bullying and harassment?	☐	☐
14.3 Are disciplinary measures in force to cover these issues?	☐	☐
14.4 Are employees exposed to the risk of violence, bullying and harassment when dealing with clients, customers and members of the public?	☐	☐

(*Continued*)

Offices and commercial premises (*Continued*)

15. Procedures	Yes	No
15.1 Are the premises inspected on a regular basis with the intention of identifying hazards and specifying measures to eliminate or control these hazards?	☐	☐
15.2 Are risk assessments undertaken on a regular basis with respect to:		
• the workplace?	☐	☐
• work equipment?	☐	☐
• work activities?	☐	☐
• personal protective equipment?	☐	☐
• manual handling operations?	☐	☐
• display screen equipment?	☐	☐
• substances hazardous to health?	☐	☐
• noise?	☐	☐
• stress at work?	☐	☐
15.3 Are the outcomes of risk assessments recorded and action taken appropriately in respect of recommendations arising from the risk assessment process?	☐	☐

Personal protective equipment

	Yes	No
1.1 Is the PPE suitable:		
• in terms of preventing or controlling exposure to risk?	☐	☐
• for the work being undertaken?	☐	☐
1.2 Are the needs of the user taken into account in terms of:		
• comfort?	☐	☐
• ease of movement?	☐	☐
• convenience in putting on, use and removal?	☐	☐
• individual suitability?	☐	☐
• vision?	☐	☐
• perception of hazards?	☐	☐
1.3 Are the ergonomic requirements and state of health of persons using the PPE taken into account?	☐	☐
1.4 Is the PPE capable of fitting the wearer correctly, if necessary after adjustments?	☐	☐
1.5 Is the PPE compatible with other PPE worn?	☐	☐
1.6 Have the risks involved at the place where exposure may occur been assessed?	☐	☐
1.7 Is the PPE appropriate in protecting operators against the identified risks?	☐	☐
1.8 Has the scale of the hazard been identified?	☐	☐
1.9 Have standards representing recognised 'safe limits' for the hazard been taken into account?	☐	☐
1.10 Are there specific requirements under regulations?	☐	☐
1.11 Are there specific job requirements or restrictions which must be considered?	☐	☐
1.12 Have environmental stressors present been taken into account?	☐	☐
1.13 Is the PPE easy to clean, sanitise and maintain and is replacement of parts simple?	☐	☐

(Continued)

Personal protective equipment (*Continued*)

	Yes	No
1.14 Is accommodation provided for PPE when not being used?	☐	☐
1.15 Is information, instruction and training provided for all users of the PPE?	☐	☐
1.16 Are steps taken by managers to ensure correct use of the PPE?	☐	☐
1.17 Are there arrangements for employees to report loss of or defect in PPE?	☐	☐

Radiation hazards

		Yes	No
1.1	Are safe systems of work, which restrict exposure, prepared and in use?	☐	☐
1.2	Does prior assessment of radiation risks take place?	☐	☐
1.3	Is there a contingency plan in operation in respect of foreseeable accident, occurrence or incident?	☐	☐
1.4	Is there adequate control over sealed sources and unsealed radioactive substances held in the hand or directly manipulated by hand?	☐	☐
1.5	Is there adequate management of controlled areas and supervised areas?	☐	☐
1.6	Is there a system for formal designation of classified persons for working in controlled areas?	☐	☐
1.7	Is radiation monitoring of controlled or supervised areas undertaken?	☐	☐
1.8	Do operators receive pre-employment and continuing medical examinations?	☐	☐
1.9	Are competent radiation protection advisers appointed?	☐	☐
1.10	Are radiation protection supervisors appointed?	☐	☐
1.11	Is there a system for authorisation of persons prior to use of accelerators or X-ray sets for specified purposes?	☐	☐
1.12	Is information, instruction and training provided for radiation protection advisers and employees?	☐	☐
1.13	Are written local rules established and brought to the attention of persons involved?	☐	☐
1.14	Is appropriate supervision provided at all times?	☐	☐
1.15	Is there a formal system of dosimetry and maintenance of dose records?	☐	☐

(Continued)

Radiation hazards (*Continued*)

	Yes	No
1.16 Is there a system for training and maintaining employee awareness?	☐	☐
1.17 Is continuous and spot-check radiation (dose) monitoring undertaken?	☐	☐
1.18 Does the employer adhere to maximum permissible dose limits?	☐	☐
1.19 Are all sources adequately enclosed?	☐	☐
1.20 Is there efficient ventilation of working area?	☐	☐
1.21 Are impervious working surfaces provided and maintained?	☐	☐
1.22 Are all work techniques controlled and supervised?	☐	☐
1.23 Are remote control handling facilities provided?	☐	☐
1.24 Are there adequate arrangements for the control, accounting for, safe keeping, safe transporting and moving of radioactive substances?	☐	☐
1.25 Are there adequate arrangements for ensuring safety of equipment used for medical exposure?	☐	☐
1.26 Is there formal investigation of exposure to ionising radiation:		
• if three tenths of the annual whole body dose is exceeded?	☐	☐
• where a person has been overexposed as a result of malfunction or defect in equipment?	☐	☐
1.27 Is there a procedure for notification of occurrences to the HSE?	☐	☐
1.28 Are washing and clothing changing facilities adequate and suitable?	☐	☐
1.29 Is appropriate personal protective equipment provided and used?	☐	☐

Violence at work – personal risk checklist

		Yes	No
1.1	Do you ensure that someone knows where you are at any particular point in time?	☐	☐
1.2	If you change your plans, do you inform someone in authority?	☐	☐
1.3	Do you check or vet people you go to meet alone?	☐	☐
1.4	Do you ensure that you can be contacted at all times?	☐	☐
1.5	Where there is a check-in system, do you use it?	☐	☐
1.6	Do you consider where you park, ensuring that it's safe and well-lit?	☐	☐
1.7	Do you ensure that you use the safest route, not necessarily the quickest?	☐	☐
1.8	Do you limit the amount of money and other valuable items that you carry?	☐	☐
1.9	Do you take appropriate precautions if you are alone at work at any time?	☐	☐
1.10	Do you ensure that you are properly protected from members of the public?	☐	☐
1.11	Do you carry a personal alarm?	☐	☐

Work equipment

		Yes	No
1.1	Is the equipment suitable for the purpose for which it is to be used?	☐	☐
1.2	Is it maintained in an efficient state, in efficient working order and in good repair?	☐	☐
1.3	Where there may be specific risks from use of work equipment, is use restricted to designated persons?	☐	☐
1.4	Is there adequate information and, where appropriate, written instructions in use of work equipment?	☐	☐
1.5	Are persons using specific items of work equipment trained in the use of same?	☐	☐
1.6	Are there prescribed precautions with respect to dangerous parts of work equipment?	☐	☐
1.7	Are measures taken to prevent or control specified hazards from use of work equipment?	☐	☐
1.8	Is there adequate protection in the case of equipment working at very high or low temperature?	☐	☐
1.9	Are the controls for starting or making a significant change in operating conditions adequate with respect to:		
	• stop controls?	☐	☐
	• emergency stop control?	☐	☐
	• general controls?	☐	☐
	• control systems?	☐	☐
	• means of isolation from sources of energy?	☐	☐
1.10	Is adequate lighting provided?	☐	☐
1.11	Are maintenance operations undertaken safely?	☐	☐
1.12	Are hazards and danger points clearly marked?	☐	☐
1.13	Are there adequate warnings and warning devices?	☐	☐

Workplaces

1. Maintenance	Yes	No
1.1 Is the workplace, equipment, devices and systems maintained in an efficient state, in efficient working order and in good repair?	☐	☐
1.2 Are equipment, devices and systems, where appropriate, under a suitable system of maintenance?	☐	☐
1.3 Are mechanical ventilation systems subject to regular maintenance?	☐	☐
2. Environmental factors		
2.1 Is there effective and suitable provision for workplace ventilation?	☐	☐
2.2 Is ventilation plant equipped with warning devices to indicate plant failure?	☐	☐
2.3 Is a reasonable temperature maintained at all times?	☐	☐
2.4 Are heating and cooling systems safe and with no emission of gases, fumes, etc?	☐	☐
2.5 Is a sufficient number of thermometers provided?	☐	☐
2.6 Is suitable and sufficient lighting installed?	☐	☐
2.7 Is suitable and sufficient emergency lighting installed?	☐	☐
2.8 Is the cleanliness of: • workplace, furniture, furnishings and fittings • floor, wall and ceiling surfaces maintained to a satisfactory level?	☐ ☐	☐ ☐
2.9 Are accumulations of waste materials removed regularly and suitable receptacles provided for waste?	☐	☐
2.10 Is sufficient working space provided for all employees?	☐	☐

(*Continued*)

Workplaces (*Continued*)

		Yes	No
2.11	Are workstations suitable for persons using them and for the work being undertaken?	☐	☐
2.12	Are suitable seats, including footrests, provided at workstations?	☐	☐

3. Structure

		Yes	No
3.1	Are all floors and traffic routes suitable for purpose of use?	☐	☐
3.2	Are measures taken to ensure there are no dangerous holes, slopes and uneven or slippery surfaces?	☐	☐
3.3	Is effective means of floor drainage installed, where necessary?	☐	☐
3.4	Are all floors and traffic routes kept free from obstruction?	☐	☐
3.5	Are hand rails and guards provided on staircases?	☐	☐
3.6	Are measures taken to prevent: • a person falling a distance likely to cause injury?	☐	☐
	• any person being struck by a falling object?	☐	☐
3.7	Are areas where above risks could arise suitably marked?	☐	☐
3.8	Are tanks, pits and other structures, where a person could fall into a dangerous substance, securely covered or fenced?	☐	☐
3.9	Are traffic routes over, across or in an uncovered tank, pit or structure securely fenced?	☐	☐
3.10	Are windows, transparent or translucent surfaces in walls, partitions, doors and gates of safety material and suitably marked?	☐	☐
3.11	Do all windows, skylights and ventilators open in a safe manner?	☐	☐

(*Continued*)

Workplaces (*Continued*)

		Yes	No
3.12	Are windows and skylights designed or constructed to ensure safe cleaning?	☐	☐
3.13	Are devices installed to enable the safe cleaning of windows and skylights?	☐	☐
3.14	Can pedestrians and vehicles circulate in a safe manner?	☐	☐
3.15	Are all traffic routes: • suitable for the persons or vehicles using them? • sufficient in number? • in suitable positions? • of sufficient size?	☐ ☐ ☐ ☐	☐ ☐ ☐ ☐
3.16	Are all traffic routes suitably indicated?	☐	☐
3.17	All doors and gates suitably constructed and fitted with any necessary safety devices?	☐	☐
3.18	Are all escalators and moving walkways: • functioning safely? • equipped with necessary safety devices? • fitted with easily identifiable and readily accessible stop controls?	☐ ☐ ☐	☐ ☐ ☐
4.	**Welfare amenity provisions**		
4.1	Are suitable and sufficient sanitary conveniences provided at readily accessible places?	☐	☐
4.2	Are sanitary convenience areas: • adequately ventilated and well lit? • kept in a clean and orderly condition?	☐ ☐	☐ ☐
4.3	Are separate sanitary convenience areas provided for men and women?	☐	☐
4.4	Are suitable and sufficient washing facilities, including showers, provided at readily accessible places?	☐	☐

(*Continued*)

Workplaces (*Continued*)

	Yes	No
4.5 Are washing facilities:		
• provided in the immediate vicinity of every sanitary convenience?	☐	☐
• provided in the vicinity of changing rooms?	☐	☐
• provided with a supply of clean hot and cold water, or warm water?	☐	☐
• provided with soap and other means of cleaning?	☐	☐
• provided with towels or other suitable means of drying?	☐	☐
• sufficiently ventilated and well lit?	☐	☐
• kept in a clean and orderly condition?	☐	☐
4.6 Are separate washing facilities provided for men and women?	☐	☐
4.7 Is an adequate supply of drinking water provided and:		
• readily accessible at suitable places?	☐	☐
• conspicuously marked?	☐	☐
4.8 Are cups or drinking vessels provided, unless the water supply is in a jet?	☐	☐
4.9 Is suitable and sufficient accommodation provided:		
• for clothing not worn during working hours?	☐	☐
• for special clothing worn only at work?	☐	☐
4.10 Does the clothing accommodation:		
• provide suitable security for clothes not worn at work?	☐	☐
• include, where necessary, separate accommodation for clothing worn at work and for other clothing?	☐	☐
• include facilities for drying clothing?	☐	☐
4.11 Is clothing accommodation in a suitable location?	☐	☐
4.12 Are suitable and sufficient facilities provided for changing clothing:		
• where workers have to wear special clothing for work?	☐	☐

(*Continued*)

Workplaces (*Continued*)

	Yes	No
• where, for reasons of health or propriety, workers cannot be expected to change elsewhere?	☐	☐
4.13 Are separate facilities, or the separate use of facilities, provided for men and women?	☐	☐
4.14 Are suitable and sufficient rest facilities provided at readily accessible places?	☐	☐
4.15 Do rest facilities include suitable facilities to eat meals where food eaten in the workplace would be likely to become contaminated?	☐	☐
4.16 Do rest facilities include suitable arrangements to protect non-smokers from discomfort caused by tobacco smoke?	☐	☐
4.17 Are suitable rest facilities provided for pregnant women and nursing mothers?	☐	☐
4.18 Are suitable and sufficient facilities provided for persons at work to eat meals?	☐	☐

PART 3
Health and Safety Information

3(a)
Tables and figures

Accident indices.

$$\text{Frequency rate} = \frac{\text{Total number of accidents}}{\text{Total number of man-hours worked}} \times 10\,000$$

$$\text{Incidence rate} = \frac{\text{Total number of accidents}}{\text{Number of persons employed}} \times 1000$$

$$\text{Severity rate} = \frac{\text{Total number of days lost}}{\text{Total number of man-hours worked}} \times 1000$$

$$\text{Mean duration rate} = \frac{\text{Total number of days lost}}{\text{Total number of accidents}}$$

$$\text{Duration rate} = \frac{\text{Number of man-hours worked}}{\text{Total number of accidents}}$$

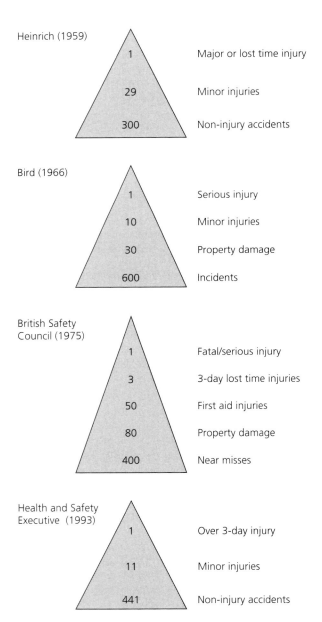

Accident ratios.

Airborne contaminants: comparison of particle size ranges.

Contaminant	Particle size range (microns)	Characteristics
Dust	0.1–75.0	Generated by natural fragmentation or mechanical cutting or crushing of solids e.g. wood, rock, coal, metals, etc. *Grit* particles, usually to be considered above 75 microns, are unlikely to remain airborne
Fume	0.001–1.0	Small solid particles of condensed vapour, especially metals, as in welding or metal melting processes. Often agglomerate into larger particles as the smaller particles collide
Smoke	0.01–1.0	Aerosol formed by incomplete combustion of organic matter; does not include ash e.g. fly ash
Mist	0.01–10.0	Aerosol of droplets formed by condensation from the gaseous state or as dispersion of a liquid state e.g. hot open surface tank processes, electroplating
Vapour	0.005	Gaseous state of materials that are liquid or solid at normal room temperature and pressure e.g. solvent vapours
Gas	0.0005	Materials which do not normally exist as liquids or solids at normal room temperature and pressure

Air changes per hour (comfort ventilation).

Location	Summer	Winter
Offices	6	4
Corridors	4	2
Amenity areas	6	4
Storage areas	2	2
Production areas (with heat-producing plant)	20	20
Production areas (assembly, finishing work)	6	6
Workshops	6	4

Average illuminances and minimum measured illuminances.

General activity	Typical locations/types of work	Average illuminance (lux (Lx))	Minimum measured illuminance (lux (Lx))
Movement of people, machines and vehicles (1)	Lorry parks, corridors circulation routes	20	5
Movement of people, machines and vehicles in hazardous areas; rough work not requiring any perception of detail (1)	Construction site clearance excavation and soil work, docks, loading bays, bottling and canning plants	50	20
Work requiring limited perception of detail (2)	Kitchens, factories assembling large components, potteries	100	50
Work requiring perception of detail (2)	Offices, sheet metal work, bookbinding	200	100
Work requiring perception of fine detail (2)	Drawing offices, factories assembling electronic components, textile production	500	200

Categories of danger – Chemicals (Hazard Information and Packaging for Supply) Regulations 2002.

Category	Property (see Note 1)	Symbol letter
Physico-chemical properties		
Explosive	Solid, liquid, pasty or gelatinous substances and preparations which may react exothermically without atmospheric oxygen thereby quickly evolving gases, and which under defined test conditions detonate, quickly deflagrate or upon heating explode when partially confined	E
Oxidising	Substances and preparations which give rise to an exothermic reaction in contact with other substances, particularly flammable substances	O
Extremely flammable	Liquid substances and preparations having an extremely low flash point and a low boiling point and gaseous substances and preparations which are flammable in contact with air at ambient temperature and pressure	F+
Highly flammable	The following substances and preparations, namely: (a) substances and preparations which may become hot and finally catch fire in contact with air at ambient temperature without any application of energy (b) solid substances and preparations which may readily catch fire after brief contact with a source of ignition and which continue to burn or to be consumed after removal of the source of ignition (c) liquid substances and preparations having a very low flash point (d) substances and preparations which, in contact with water or damp air, evolve highly flammable gases in dangerous quantities (see Note 2)	F

(*Continued*)

Categories of danger (*Continued*)

Category	Property (see Note 1)	Symbol letter
Flammable	Liquid substances and preparations having a low flash point	None
Health effects		
Very toxic	Substances and preparations which *in very low quantities* can cause death or acute or chronic damage to health when inhaled, swallowed or absorbed via the skin	T+
Toxic	Substances and preparations which in *low quantities* can cause death or acute or chronic damage to health when inhaled, swallowed or absorbed via the skin	T
Harmful	Substances and preparations which may cause death or acute or chronic damage to health when inhaled, swallowed or absorbed via the skin	Xn
Corrosive	Substances and preparations which may, on contact with living tissue, destroy them	C
Irritant	Non-corrosive substances and preparations which through immediate, prolonged or repeated contact with the skin or mucous membrane, may cause inflammation	Xi
Sensitising	Substances and preparations which, if they are inhaled or if they penetrate the skin, are capable of eliciting a reaction by hypersensitisation such that on further exposure to the substance or preparation, characteristic adverse effects are produced	
Sensitising by inhalation		Xn
Sensitising by skin contact		Xi
Carcinogenic (see Note 3)	Substances and preparations which, if they are inhaled or ingested or if they	

(*Continued*)

Category	Property (see Note 1)	Symbol letter
	penetrate the skin, may induce cancer or increase its incidence	
Category 1		T
Category 2		T
Category 3		Xn
Mutagenic (see Note 3)	Substances and preparations which, if they are inhaled or ingested or if they penetrate the skin, may induce heritable genetic defects or increase their incidence	
Category 1		T
Category 2		T
Category 3		Xn
Toxic for reproduction	Substances and preparations which, if they are inhaled or ingested or if they penetrate the skin, may produce or increase the incidence of non-heritable adverse effects in the progeny and/or impairment of male or female reproductive functions or capacity	
Category 1		T
Category 2		T
Category 3		Xn
Dangerous for the environment (see Note 4)	Substances which, were they to enter into the environment, would or might present an immediate or delayed danger for one or more components of the environment	N

Notes
1. As further described in the *approved classification and labelling guide* (p. 47)
2. Preparations packed in *aerosol dispensers* shall be classified as *flammable* in accordance with the additional criteria set out in Part II of this Schedule
3. The categories are specified in the *approved classification and labelling guide*
4. (a) In certain cases specified in the *approved supply list* and in the *approved classification and labelling guide* as substances classified as *dangerous for the environment* do not require to be labelled with the symbol for this category of danger
   (b) This category of danger does not apply to preparations.

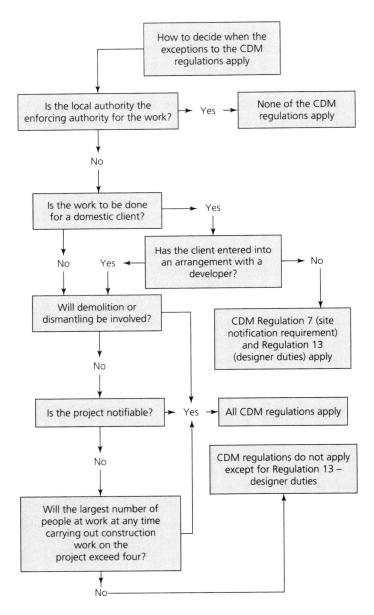

Construction (Design and Management) Regulations 1994 – How to decide when the exceptions to the CDM regulations apply.

Decibels (addition of).

Difference (dB)	Add to higher (dB)
0–0.5	3.0
1.0–1.5	2.5
2.0–3.0	2.0
3.5–4.5	1.5
5.0–7.0	1.0
7.5–12.0	0.5
Over 12.0	0

Demolition methods (Demolition – BS 6187: 1982 A guide to typical methods of demolition).

Type of structure	Type of construction	Method of demolition				
		Detached building isolated site	Detached building confined site	Attached building isolated site	Attached building confined site
Small and medium two-storey building	Loadbearing walls	ABCDM	ABDM	ABDM	ADM
Large buildings three storeys and over	Loadbearing walls	ABDM	ABDM	ABDM	AD
	Loadbearing walls with wrought iron and cast iron members	ABDM	AM	AM	AM
Framed structures	Structural steel	ACM	AM	AM	AM
	In situ reinforced concrete	ADM	ADM	ADM	AM
	Pre-cast reinforced concrete	ADM	ADM	ADM	AM
	Prestressed reinforced concrete	ADM	ADM	ADM	AM
	Composite (structural steel and reinforced concrete)	ADM	ADM	ADM	AM
Independent cantilevers (canopies, balconies and staircases)	Timber	ABCDM	ABDM	ABDM	ABDM
		ADM	ADM	ADM	ADM

Structure				
Bridges	ABCDM	ABCDM	AM	AM
Masonry arches	ACDM	ACDM	ACDM	ACDM
Chimneys — Brick or masonry	ACD	A	ACD	A
Chimneys — Steel	AC	A	A	A
Chimneys — In situ and precast reinforced concrete	AD	A	A	A
Chimneys — Reinforced plastics				
Spires	AC	A	A	A
Pylons and masts	ACD	A	A	A
Petroleum tanks (underground)	AC	A		
Above ground storage tanks				
Chemical works and similar establishments				
Basements				
Special structures				

Note 1

This table is a general guide to the methods of demolition usually adopted in particular circumstances. In addition, subject to local restraints, explosives may be used by experienced personnel in many of the circumstances listed. The indication of a particular method does not necessarily preclude the use of another method, or the use of several methods in combination.

Note 2 Legend

A: hand demolition; B: mechanical demolition by pusher; C: mechanical demolition by deliberate collapse; D: mechanical demolition by demolition ball; M: demolition by other mechanical means excluding wire pulling.

Electromagnetic spectrum.

Radiation	Frequency	Wavelength	Energy	Radiation sources
Gamma	10^{21}	Short	High	Cosmic sources
X-ray	10^{18}			Atoms struck by high energy particles
Ultraviolet light				Excited gases
Visible light	10^{15}			Hot bodies
Infrared	10^{12}–10^{14}			Hot bodies
Microwaves	10^{9}			Microwave
Radio waves	-10^{6}	Long	Low	Radio transmitter

Fire instruction notice.

When the fire alarm sounds
1. Close the windows, switch off electrical equipment and leave the room closing the door behind you.
2. Walk quickly along the escape route to the open air.
3. Report to the fire warden at your assembly point.
4. Do not attempt to re-enter the building.

When you find a fire
1. Raise the alarm by. . . . .
   (If the telephone is used, the notice must include a reference to name and location.)
2. Leave the room, closing the door behind you.
3. Leave the building by the escape route.
4. Report to the fire warden at the assembly point.
5. Do not attempt to re-enter the building.

Hazardous substances that can be revealed by medical analysis.

Substance	Technique
Benzene	Phenol in urine; benzene in breath
Inorganic lead	Lead in blood/urine; coproporphyrin in urine
Elemental mercury/ inorganic mercury	Mercury in urine; protein in urine
Methyl mercury	In faeces
Arsenic	In urine, hair, nails
Calcium	In blood, urine
Trichlorethylene	In urine as trichloracetic acid
Organo-phosphorus compounds	Cholinesterase in blood/urine; nerve conduction velocity; electromyography

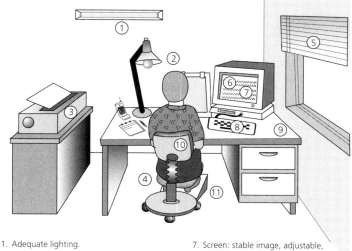

1. Adequate lighting.
2. Adequate contrast, no glare or distracting reflections.
3. Distracting noise minimised.
4. Legroom clearances to allow postural changes.
5. Window covering.
6. Software: appropriate to task, adapted to user, provides feedback on system status, no undisclosed monitoring.
7. Screen: stable image, adjustable, readable, glare/reflection free.
8. Keyboard: useable, adjustable, detachable, legible.
9. Work surface: allow flexible arrangements, spacious, flare-free.
10. Work chair: adjustable.
11. Foot rest.

Health and Safety (Display Screen Equipment) Regulations 1992. (a) Display screen equipment workstation – design and layout.

1. Seat back adjustability.
2. Good lumbar support.
3. Seat height adjustability.
4. No excess pressure on underside of thighs and back of knees.
5. Foot supported if needed.
6. Space for postural change, no obstacles under desk.
7. Forearms approximately horizontal.
8. Minimal extension, flexion or deviation of wrists.
9. Screen height and angle should allow comfortable head position.
10. Space in front of keyboard to support hands/wrists during pauses in keying

Health and Safety (Display Screen Equipment) Regulations 1992.
(b) Seating and posture for typical office tasks.

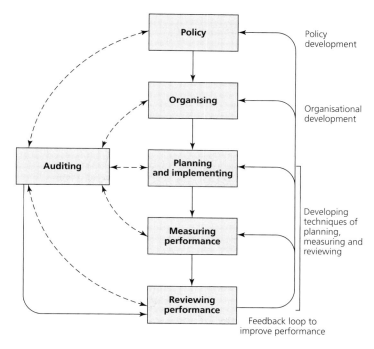

Key elements of successful health and safety management.

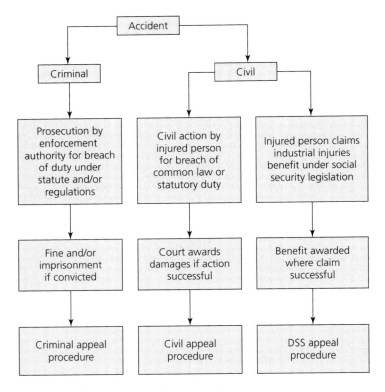

Legal routes following an accident at work.

Local exhaust ventilation (LEV) systems.

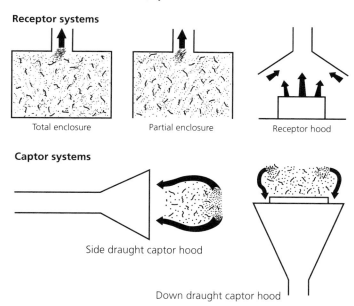

Receptor systems

Total enclosure

Partial enclosure

Receptor hood

Captor systems

Side draught captor hood

Down draught captor hood

Typical LEV system (Woodcutting machinery)

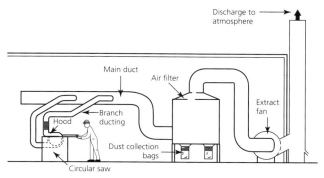

Discharge to atmosphere

Main duct

Air filter

Extract fan

Branch ducting

Hood

Dust collection bags

Circular saw

If the hands enter MORE THAN ONE of the box zones during the operation, the SMALLEST weight figure should be used.

The transition from one box zone to another is not abrupt; an INTERMEDIATE figure may be chosen where the hands are close to a boundary.

Where lifting or lowering with the hands BEYOND the box zones is UNAVOID-ABLE, a more detailed assessment should be made.

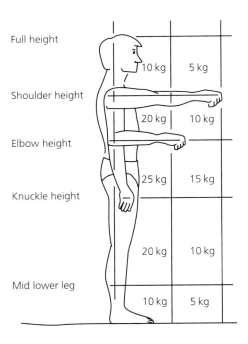

Manual handling – lifting and lowering.

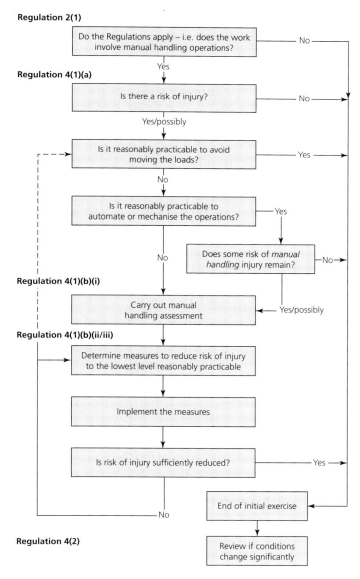

Manual handling operations regulations – flow chart.

Maximum permissible wet globe temperatures.

Work:rest schedule (per hour)	Work load		
	Light	Moderate	Heavy
Continuous work	30.0°C	26.7°C	25.0°C
75% work:25% rest	30.6°C	28.0°C	25.9°C
50% work:50% rest	31.4°C	29.4°C	27.9°C
25% work:75% rest	32.2°C	31.1°C	30.0°C

Maximum ratios of illuminance for adjacent areas.

Situation to which recommendation applies	Typical location	Maximum ratio of illuminances		
		Working area		Adjacent area
Where each task is individually lit and the area around the task is lit to a lower illuminance	Local lighting in an office	5	:	1
Where two working areas are adjacent, but one is lit to a lower illuminance than the other	Localised lighting in a works store	5	:	1
Where two working areas are lit to different illuminances but are separated by a barrier and there is frequent movement between them	A storage area inside a factory and a loading bay outside	10	:	1

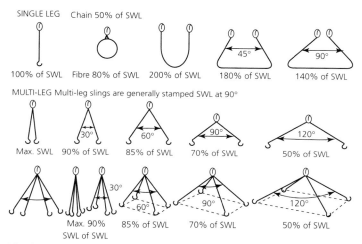

SINGLE LEG   Chain 50% of SWL

100% of SWL   Fibre 80% of SWL   200% of SWL   180% of SWL (45°)   140% of SWL (90°)

MULTI-LEG Multi-leg slings are generally stamped SWL at 90°

Max. SWL   90% of SWL (30°)   85% of SWL (60°)   70% of SWL (90°)   50% of SWL (120°)

Max. 90% SWL of SWL (30°/60°)   85% of SWL   70% of SWL (90°)   50% of SWL (120°)

Select the correct size of a sling for the load taking into account the included angle and the possibility of unequal loading in the case of multi-leg slings

Maximum safe working loads for slings at various angles.

Noise control methods.

Sources and pathways	Control measures
Vibration produced through machinery operation	Reduction at source
Structure-borne noise (vibration)	Vibration isolation, e.g. resilient mounts and connections, anti-vibration mounts
Radiation of structural vibration	Vibration damping to prevent resonance
Turbulence created by air or gas flow	Reduction at source or use of silencers
Airborne noise pathway	Noise insulation – reflection; heavy barriers
	Noise absorption – no reflection; porous lightweight barriers

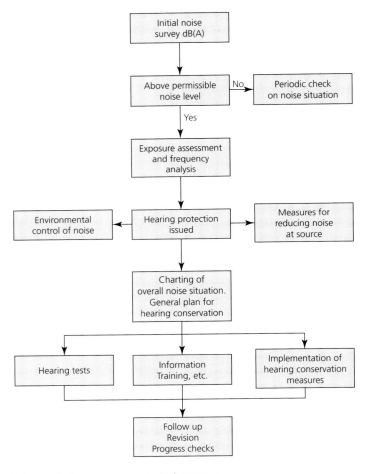

Noise control programme – typical structure.

Octave bands (standard range).

Limits of band (Hz)	Geometric centre frequency (Hz)
45–90	63
90–180	125
180–355	250
355–710	500
710–1400	1000
1400–2800	2000
2800–5600	4000
5600–11200	8000

Optimum working temperatures.

Type of work	Optimum temperatures (°C)
Sedentary/office work	
Comfort range	19.4–22.8
Light work	
Optimum temperature	18.3
Comfort range	15.5–20.0
Heavy work	
Comfort range	12.8–15.6

Personal Protective Equipment at Work Regulations 1992 – Specimen risk survey table for the use of personal protective equipment.

The PPE at Work Regulations 1992 apply except where the Construction (Head) Protection Regulations 1989 apply

The CLW, IRR, CAW, COSHH and NAW Regulations[1] will each apply to the appropriate hazard

			Mechanical					Thermal			Non-ionising radiation	Electrical	Noise	Ionising radiation	Dust fibre	Fume	Vapours	Splashes, spurts	Gases, vapours	Harmful bacteria	Harmful viruses	Fungi	Non-micro biological antigens
			Falls from a height	Blows, cuts, impact, crushing	Stabs, cuts, grazes	Vibration	Slipping, falling over	Scalds, heat, fire	Cold	Immersion													
P		Cranium																					
A	Head	Ears																					
R		Eyes																					
T		Respiratory tract																					
S		Face																					
		Whole head																					
of	Upper limbs	Hands																					
the		Arms (parts)																					
B	Lower limbs	Foot																					
O		Legs (parts)																					
	Various	Skin																					
D		Trunk/abdomen																					
Y		Whole body																					

(1) The Control of Lead at Work Regulations 1980, The Ionising Radiations Regulations 1985, The Control of Asbestos at Work Regulations 1987, The Control of Substances Hazardous to Health Regulations 1988, The Noise at Work Regulations 1989.

Places of work requiring inspection by a competent person under (Regulation 29(1) of the Construction (Health, Safety and Welfare) Regulations 1996.

Place of work	Time of inspection
1. Any working platform or part thereof or any personal suspension equipment provided pursuant to paragraph 3(b) or (c) of Regulation 6	1. (i) Before being taken into use for the first time; and (ii) after any substantial addition, dismantling or other alteration; and (iii) after any event likely to have affected its strength or stability; and (iv) at regular intervals not exceeding seven days since the last inspection
2. Any excavation which is supported pursuant to paragraphs 1, 2 or 3 of Regulation 12	2. (i) Before any person carries out work at the start of every shift; and (ii) after any event likely to have affected the strength or stability of the excavation or any part thereof; and (iii) after any accidental fall of rock or earth or other material
3. Cofferdams and caissons	3. (i) Before any person carries out work at the start of every shift; and (ii) after any event likely to have affected the strength or stability of the cofferdam or caisson or any part thereof

Probability index.

Probability index	Descriptive phrase
10	Inevitable
9	Almost certain
8	Very likely
7	Probable
6	More than even chance
5	Even chance
4	Less than even chance
3	Improbable
2	Very improbable
1	Almost impossible

Reporting of Injuries, Diseases and Dangerous Occurrences Regulations 1995 – reporting requirements.

Reportable event	Person affected	Responsible person
1. Special cases		
All reportable events in mines		The mine manager
All reportable events in quarries or in closed mine or quarry tips		The owner
All reportable events at offshore installations, except cases of disease reportable under Regulation 5		The owner, in respect of a mobile installation, or the operator in respect of a fixed installation (under these regulations the responsibility extends to reporting incidents at sub-sea installations, except tied back wells and adjacent pipeline)
All reportable events at diving installations, except cases of disease reportable under Regulation 5		The diving contractor
2. Injuries and disease		
Death, major injury, over 3-day injury, or case of disease connected with diving operations and work at an offshore installation	An employee at work	That person's employer
	A self-employed person at work in premises under the control of someone else	The person in control of the premises: • at the time of the event • in connection with the carrying on of any trade, business or undertaking
Major injury, over 3-day injury or case of disease	A self-employed person at work in premises under their control	The self-employed person or someone acting on his behalf

(*Continued*)

Reporting of Injuries, Diseases and Dangerous Occurrences Regulations 1995 – reporting requirements (*Continued*).

Reportable event	Person affected	Responsible person
Death or injury requiring removal to a hospital for treatment (or major injury occurring at a hospital)	A person who is not at work (but is affected by the work of someone else) e.g. a member of the public, student, a resident of a nursing home	The person in control of the premises where, or in connection with the work going on at which, the accident causing the injury happened: • at the time of the event • in connection with their carrying on of any trade, business or undertaking
3. *Dangerous occurrences* One of the dangerous occurrences listed in Schedule 2 to the regulations, except • where they occur at workplaces covered by Part 1 of this Table (i.e. mines, quarries, closed mine or quarry tips, offshore installations or connected with diving operations); or • those covered below		The person in control of the premises where, or in connection with the work going on at which, the dangerous occurrence happened: • at the time the dangerous occurrence happened, and • in connection with their carrying on trade, business or undertaking
A dangerous occurrence at a well		The concession owner (the person having the right to exploit or explore mineral resources and store and recover gas in any area, if the well is used or is to be used to exercise that right) or the person appointed by the concession owner to organise or supervise any operation carried out by the well
A dangerous occurrence at a pipeline but not a dangerous occurrence connected with pipeline works		The owner of the pipeline
A dangerous occurrence involving a dangerous substance being conveyed by road		The operator of the vehicle

Safety data sheets – obligatory headings (CHIP Regulations).

1. Identification of the substance or preparation
2. Composition/information on ingredients
3. Hazard identification
4. First Aid measures
5. Fire-fighting measures
6. Accidental release measures
7. Handling and storage
8. Exposure controls/Personal protection
9. Physical and chemical properties
10. Stability and reactivity
11. Toxicological information
12. Ecological information
13. Disposal considerations
14. Transport information
15. Regulatory information
16. Other information

Safety signs.

Meaning or purpose	Safety colour	Examples of use	Contrasting colour (if required)		Symbol
Stop Prohibition	Red	STOP signs; prohibition signs identification of emergency shutdown devices	White	Black	 No smoking
Caution Risk of danger	Yellow	Warning signs e.g. electric current on, harmful vapours, obstacle ahead, scaffold incomplete, asbestos	Black	Black	 Explosive
Mandatory action	Blue	Obligation to wear personal protective equipment e.g. eye protection; report damage immediately; keep out; switch off machine when not in use	White	White	 Ear protection must be worn
Safe condition	Green	Identification of first aid posts, safety showers, fire exits	White	White	 Means of Escape

Severity index.

Severity index	Descriptive phrase
10	Death
9	Permanent total incapacity
8	Permanent severe incapacity
7	Permanent slight incapacity
6	Absent from work for more than 3 weeks with subsequent recurring incapacity
5	Absent from work for more than 3 weeks but with subsequent complete recovery
4	Absent from work for more than 3 days but less than 3 weeks with subsequent complete recovery
3	Absent from work for less than 3 days with complete recovery
2	Minor injury with no lost time and complete recovery
1	No human injury expected

The total working system – areas of study.

Human characteristics	*Environmental factors*
Body dimensions	Temperature
Strength	Humidity
Physical and mental limitations	Light
Stamina	Ventilation
Learning	Noise
Perception	Vibration
Reaction	
Man–machine interface	*Total working system*
Displays	Fatigue
Controls	Work rate
Communications	Posture
Automation	Stress
	Productivity
	Accidents
	Safety

Water closets and urinals for men (ACOP to Workplace (Health, Safety and Welfare) Regulations 1992).

Number of men at work	Number of water closets	Number of urinals
1–15	1	1
16–30	2	1
31–45	2	2
46–60	3	2
61–75	3	3
76–90	4	3
91–100	4	4

Water closets and wash station provision (ACOP to Workplace (Health, Safety and Welfare) Regulations 1992).

Number of people at work	Number of water closets	Number of wash stations
1–5	1	1
6–25	2	2
26–50	3	3
51–75	4	4
75–100	5	5

3(b)
Forms

Accident book entry

ACCIDENT RECORD

☐ Report Number

1 About the person who had the accident

Name: ....

Address: ....

.... Postcode: ....

Occupation: ....

2 About you, the person filling in this record

▼ If you did not have the accident write your address and occupation

Name: ....

Address: ....

.... Postcode: ....

Occupation: ....

3 About the accident: *Continue on the back of this form if you need to*

▼ Say when it happened   Date: .... Time: ....

▼ Say where it happened. State which room or place ....

▼ Say how the accident happened. Give the case if you can ....

....

....

....

▼ If the person who had the accident suffered an injury, say what it was

....

▼ Please sign the record and date it

....

Signature: .... Date: ....

4 For the employer only

▼ Complete this box if the accident is reportable under the Reporting of Injuries, Diseases and Dangerous Occurrences Regulations 1995 (RIDDOR).

How was it reported?

....

Date reported   /   /   Signature

....

Construction (Design and Management) Regulations 1994 – Notification of Project (Form 10)

The Data Protection Act 1998 requires the Health and safety Executive (HSE) to inform you that this form may include information about you (this is called 'personal data' in the Act) and that we are a 'data controller' for the purposes of this Act. HSE will process the data for health, safety and environmental purposes. HSE may disclose these data to any person or organisation for the purposes for which it was collected or where the Act allows disclosure. As data subject, you have the right to ask for a copy of the data and to ask for any inaccurate data to be corrected.

Notification of construction project

Who should use this form
- Any person who needs to notify a project covered by the Construction (Design and Management) Regulations 1994 which will last longer than 30 days or 500 person days.
- The form can be used by contractors working for domestic clients. In this case only parts 4-8 and 11 need to be filled in.
- Any person sending updated information which was not available at the time of initial notification.
- Any day on which construction work is carried out (including holidays and weekends) should be counted, even if the work on that day is of short duration.
- A person day is one individual, including supervisors and specialists, carrying out construction work for one normal working shift.)

Where to send the form
- The completed form should be sent to the HSE area office covering the site where construction work is to take place.

When to send this form
- You should send this form as soon as possible after the planning supervisor is appointed to the project. In the case of work for a domestic client the form should be sent in as soon as the contractor is appointed.

---

1 Is this the initial notification of this project or are you providing additional information not previously available?

Initial notification ☐   **Additional notification** ☐

---

2 Client: name, full address, postcode and telephone number *(if more than one client, please attach details on separate sheet)*

Name:

Address:

Postcode: Telephone number:

---

3 Planning Supervisor: name, full address, postcode and telephone no.

Name:

Address:

Postcode: Telephone number:

---

4 Principal Contractor: *(or contractor when project for a domestic client)* name, full address, postcode and telephone no.

Name:

Address:

Postcode: Telephone number:

---

5 Address of site: where construction work is to be carried out

Address:

Postcode:

F10 (rev 08.04)

6 Local Authority: name of the local government district council or island council within whose district the operations are to be carried out

7 Please give your estimates on the following: Please indicate if these estimates are original   ☐ revised   ☐ *(tick relevant box)*

a. The planned date for the commencement of the construction work

b. How long the construction work is expected to take *(in weeks)*.

c. The maximum number of people carrying out construction work on site at any one time

d. The number of contractors expected to work on site

8 Construction work: give brief details of the type of construction work that will be carried out

9 Contractors: name full address and postcode or those who have been chosen to work on the project *(if required continue on a separate sheet). (Note this information is only required when it is known at the time notification is first made to HSE, An update is not required*

Declaration of planning supervisor

10 I hereby declare that --- (name of organisation) has been appointed as planning supervisor for the project

Signed by or on behalf of the organisation---------------------------- *(print name)* -------------------------------.

Date -----------------------------------.

Declaration of principal contractor

11 I hereby declare that -- (name of principal contractor) has been appointed as principal contractor for the projectl (or contractor undertaking project for domestic client)

Signed by or on behalf of the organisation---------------------------- *(print name)* -------------------------------

Date -----------------------------------.

Control of Substances Hazardous to Health Regulations 2002 – Health Risk Assessment

This health risk assessment has been undertaken taking into account the supplier's safety data provided in accordance with the Chemicals (Hazard Information and Packaging for Supply) Regulations.

Assessment No.: ....

Location: .... **Process/Activity/Use:** ....

Substance information: ....

Name of substance: .... Chemical composition: ....

Supplier: ....

Risk information

Risk classification: .... Stated occupational exposure limits

....

Toxic/Corrosive/Harmful/Irritant WEL ....
LTEL .... STEL ....

Route(s) of entry .... Acute/Chronic/Local/Systemic

Exposure situations ....

Exposure effects ....

Estimate of potential exposure Frequency of use ....

....

Quantities used .... Duration of use ....

Storage requirements

(Continued)

Health Risk Assessment (*continued*)

Air monitoring requirements and standards
First aid requirements
Health surveillance requirements
Routine disposal requirements

Procedure in the event of spillage:

1. Small scale spillage ....

   ....

2. Large scale spillage ....

   ....

Information, instruction and training requirements/arrangements

General conclusions as to risk

High/Medium/Low risk ....

Special precautions ....

....

Supervision requirements ....

....

....

....

....

Health Risk Assessment Summary

General comments as to extent of health risk

Action

1. Immediate action

2. Short term action (7 days)

3. Medium term action (3 months)

4. Long term action (12 months)

Date of reassessment: ..../......../........

Assessor: ....   **Date:** ..../......../........

Hazard report form

HAZARD REPORT

1. Report (to be completed by person reporting hazard)

Date: ....   Time: ....   Department: ....
Reported to: .... (Verbal) .... (Written)
Description of hazard (including location, plant, machinery, practice, etc.)

Signature: .... Position: ....

2. Action (to be completed by Departmental Manager/Supervisor)

Hazard verified YES/NO   Date: .... Time: ....
Remedial action (including changes in systems of work)

Action to be taken by: Name: .... Signature: ....
*Priority Rating 1 2 3 4 5   Estimated cost: ....
Completion: Date: ....
Interim precautions

Signature: .... (Departmental Manager)

3. Financial approval (to be completed by senior manager where cost
   exceeds departmental authority)

The expenditure necessary to complete the above work is approved.

Signature: .... Date: ....

4. Completion   The remedial action described above is complete.
Actual cost: ....

Date: ....   Signature: ....
                                               (Persons completing the work)

5. Safety Adviser's check   I have checked completion of the above
   work and confirm that the hazard has been eliminated/controlled

Date: ....   Signature: ....

* **Priority Ratings** 1 – immediate; 2 – 48 hours; 3 – 1 week; 4 – 1 month;
   5 – 3 months

Improvement notice

HEALTH AND SAFETY EXECUTIVE   Serial No. 1

Health and Safety at Work etc. Act 1974, Sections 21, 23 and 24

IMPROVEMENT NOTICE

Name and address (See Section 46)	To .... (a) Trading as ....
(a) Delete as necessary	(b) ....
(b) Inspector's full name	one of (c) .... of (d) ....
(c) Inspector's official designation	.... Tel. No. .... hereby give you notice that I am of the opinion that at ....
(d) Official address	(e) ....
(e) Location of premises or place and activity	you, as (a) an employer/a self employed person/a person wholly or partly in control of the premises (f) ....
(f) Other specified capacity	(a) are contravening/have contravened in circumstances that make it likely that the contravention will continue to be repeated. ....
(g) Provisions contravened	(g) .... The reasons for my said opinion are: .... and I hereby require you to remedy the said contraventions or, as the case may be, the matters occasioning them by (h) ....
(h) Date	(a) In the manner stated in the attached schedule which forms part of the notice. Signature: .... Date: .... Being an Inspector appointed by an Instrument in writing made pursuant to Section 19 of the said Act and entitled to issue this notice. (a) An Improvement notice is also being served on .... of ....
LP1	related to the matters contained in this notice.

Job safety analysis record

Job title:

Department:

Purpose of job:

Machinery and equipment used:

Materials used:

Personal protective equipment required:

Machinery safety features (where appropriate):

Intrinsic hazards:

Degree of risk:

Work organisation:

Specific tasks:

Skills required:

Influences on behaviour:

Learning method:

Operation of the safe system of work:

Supervision requirements:

Date of next review: Signed: Date:

Manual handling of loads

EXAMPLE OF AN ASSESSMENT CHECKLIST

Note: This checklist may be copied freely. It will remind you of the main points to think about while you:

- consider the risk of injury from manual handling operations
- identify steps that can remove or reduce the risk
- decide your priorities for action.

SUMMARY OF ASSESSMENT	Overall priority for remedial action: Nil/Low/Med/High*
Operations covered by this assessment:	Remedial action to be taken: ....
....	....
....	....
Locations:	Date by which action is to be taken: ....
Personnel involved:	Date for reassessment: ....
Date of assessment:	Assessor's name: .... Signature: ....

*circle as appropriate

SECTION A – Preliminary:

Q1 Do the operations involve a significant risk of injury?   Yes/No*

If **'Yes'** go to Q2. If **'No'** the assessment need go no further.

If in doubt answer **'Yes'**. You may find the guidelines in Appendix 1 helpful.

Q2 Can the operations be avoided/mechanised/automated at reasonable cost?   Yes/No*

If **'No'** go to Q3. If **'Yes'** proceed and then check that the result is satisfactory.

Q3 Are the operations clearly within the guidelines in Appendix 1?   Yes/No*

If **'No'** go to Section B. If **'Yes'** you may go straight to Section C if you wish.

SECTION C – Overall assessment of risk:

Q   What is your overall assessment of the risk of injury?   Insignificant/Low/Med/High*

If not **'Insignificant'** go to Section D. If **'Insignificant'** the assessment need go no further.

SECTION D – Remedial action:

Q   What remedial steps should be taken, in order of priority?

i ....

ii ....

iii ....

iv ....

v ....

And finally:

- complete the SUMMARY above
- compare it with your other manual handling assessments
- decide your priorities for action
- **TAKE ACTION .... AND CHECK THAT IT HAS THE DESIRED EFFECT**

(Continued)

Manual handling of loads (*Continued*)

SECTION B – More detailed assessment, where necessary:					
Questions to consider: (If the answer to a question is 'Yes' place a tick against it and then consider the level of risk)	**Level of risk:** (Tick as appropriate)				**Possible remedial action:** (Make rough notes in this column in preparation for completing Section D)
	YES	LOW	MED	HIGH	
The tasks – do they involve:					
◆ holding loads away from trunk?	☐	☐	☐	☐	
◆ twisting?	☐	☐	☐	☐	
◆ stooping?	☐	☐	☐	☐	
◆ reaching upwards?	☐	☐	☐	☐	
◆ large vertical movement?	☐	☐	☐	☐	
◆ long carrying distances?	☐	☐	☐	☐	
◆ strenuous pushing or pulling?	☐	☐	☐	☐	
◆ unpredictable movement of loads?	☐	☐	☐	☐	
◆ repetitive handling?	☐	☐	☐	☐	
◆ insufficient rest or recovery?	☐	☐	☐	☐	
◆ a workrate imposed by a process?	☐	☐	☐	☐	
The loads – are they:					
◆ heavy?	☐	☐	☐	☐	
◆ bulky/unwieldy?	☐	☐	☐	☐	
◆ difficult to grasp?	☐	☐	☐	☐	
◆ unstable/unpredictable?	☐	☐	☐	☐	
◆ intrinsically harmful (e.g. sharp/hot?)	☐	☐	☐	☐	
The working environment – are there:					
◆ constraints on posture?	☐	☐	☐	☐	
◆ poor floors?	☐	☐	☐	☐	
◆ variations in levels?	☐	☐	☐	☐	
◆ hot/cold/humid conditions?	☐	☐	☐	☐	
◆ strong air movements?	☐	☐	☐	☐	
◆ poor lighting conditions?	☐	☐	☐	☐	
Individual capability – does the job:					
◆ require unusual capability?	☐	☐	☐	☐	
◆ hazard those with a health problem?	☐	☐	☐	☐	
◆ hazard those who are pregnant?	☐	☐	☐	☐	
◆ call for special information/training?	☐	☐	☐	☐	
Other factors:					
Is movement or posture hindered by clothing or personal protective equipment?	☐	☐	☐	☐	

Deciding the level of risk will inevitably call for judgement. The guidelines in Appendix 1 may provide a useful yardstick.

When you have completed Section B go to Section C.

Noise exposure record

Name and address of premises, department, etc.

....

....

....

Date of survey: ....   Survey made by: ....

Workplace Number of persons exposed	Noise level (Leq(s) or sound level)	Daily exposure period	$L_{EP}d$ dB(A)	Peak pressure (where appropriate)	Comments/ Remarks

General comments: ....

Instruments used: ....

....

Date of last calibration: ....

Signature: ....   Date: ....

(Noise at Work: Guide No.3: *Noise assessment, information and control*: HMSO, London)

Occupational health: (a) Pre-employment health questionnaire

Surname: .... **Forename:** ....

Date of Birth: ....

Address: ....

....

Tel. No: ....

Occupation: ....

Position applied for: ....

Name and address of doctor: ....

....

....

SECTION A

Please tick if you are at present suffering from, or have suffered from:

1. Giddiness	☐	8. Stroke	☐
Fainting attacks	☐	Heart trouble	☐
Epilepsy	☐	High blood pressure	☐
Fits or blackouts	☐	Varicose veins	☐
2. Mental illness	☐	9. Diabetes	☐
Anxiety or depression	☐	10. Skin trouble	☐
3. Recurring headaches	☐	11. Ear trouble or deafness	☐
4. Serious injury	☐	12. Eye trouble	☐
Serious operations	☐	Defective vision (not	☐
5. Severe hay fever	☐	corrected by glasses or	
Asthma	☐	contact lenses)	
Recurring chest disease	☐	Defective colour vision	☐
6. Recurring stomach trouble	☐	13. Back trouble	☐
Recurring bowel trouble	☐	Muscle or joint trouble	☐
7. Recurring bladder trouble	☐	14. Hernia/rupture	☐

(Continued)

Occupational health: (a) Pre-employment health questionnaire (*Continued*)

SECTION B

Please tick if you have any disabilities that affect:

Standing	☐	Lifting	☐	Working at heights	☐
Walking	☐	Use of your hands	☐	Climbing ladders	☐
Stair climbing	☐	Driving a vehicle	☐	Working on staging	☐

SECTION C

How many working days have you lost during the last three years due to illness or injury? .... days

Are you at present having any tablets, medicine or injections prescribed by a doctor? .... YES/NO

Are you a registered disabled person? .... YES/NO

SECTION D

Previous occupations   Duration   Name & address of employer

....

....

....

SECTION E

The answers to the above questions are accurate to the best of my knowledge.

I acknowledge that failure to disclose information may require re-assessment of my fitness and could lead to termination of employment.

Signature: .... Prospective employee   Date: ....

Signature: .... Manager   Date: ....

(*Continued*)

Occupational health (*Continued*)

ADDITIONAL QUESTIONS TO BE ANSWERED BY ANY PROSPECTIVE EMPLOYEE WHO WILL ENTER FOOD PRODUCTION AREAS OR HANDLE FOOD IN THE COURSE OF HIS/HER EMPLOYMENT

SECTION F

Please tick if you **have ever** suffered from:

Typhoid fever	☐	**A perforated**	☐	**Recurring skin**	☐
Paratyphoid	☐	**ear drum**		**condition**	
fever		**A running ear**	☐	**Hepatitis**	☐
Dysentery	☐	**Frequent**	☐	**(liver disorder)**	
Salmonella	☐	**sore throats**		**Tuberculosis**	☐

SECTION G

Please tick if you are **at present** suffering from:

Cough with	☐	**A running ear**	☐	**Diarrhoea/vomiting**	☐
phlegm		**Raised**	☐	**Boils**	☐
Abdominal pain	☐	**temperature**		**Styes**	☐
Acne	☐	**Septic fingers**	☐		

SECTION H

When did you last visit your dentist? .... 19 ....

If treatment is necessary are you willing to visit your dentist for treatment?   YES/NO

Occupational health: (b) Health questionnaire

Tick as appropriate

1. Have you ever suffered from:

Bronchitis	☐	Repeated	☐	Back pain	☐
Pleurisy	☐	sore throats		Varicose veins	☐
Tuberculosis	☐	Hernia	☐	Pneumonia	☐
Chest pain	☐	Asthma	☐	Shortness of breath	☐
Enteritis	☐	Cough	☐	Diarrhoea	☐
Typhoid	☐	Dysentery	☐	Skin disease	☐
Paratyphoid	☐	Vomiting	☐	Rashes	☐
Boils	☐	Ulcers	☐	Blackouts	☐
Hand/finger	☐	Fits	☐	Diabetes	☐
infections		Persistent	☐	Nervous disability	☐
Migraine	☐	headaches		Jaundice	☐
Ear infections	☐	Eye infections	☐	Joint pains	☐
Hay fever	☐	Allergies	☐		

2. Have you had any operations?   **YES/NO**

   If YES, state type and dates ....
   ....

3. Are you **at present** receiving any form of medical treatment?   **YES/NO**

   If YES, state form of treatment and dates ....
   ....

4. Are you taking any form of pills, medicines or drugs, prescribed or otherwise?   **YES/NO**

   If YES, state names of pills, medicines or drugs and medical reasons for taking same ....
   ....

5. When was your chest last X-rayed?   Date: ....
                                          Result: ....

6. Are you a registered disabled person?   **YES/NO**

   Reason for registration ....

7. When did you last travel abroad?

   Date: ....   Where: ....

I have answered each question to the best of my ability. I understand that deliberate misrepresentation may result in disciplinary action.

Signed: ....   Date: ....

Full name and address: ....
....

Occupational health: (c) Food handler's clearance certificate

Full name: ....

Department: .... **Clock No:** ....

Dates of absence from .... **to** ....

1. REASON FOR ABSENCE

Please tick:
(a) Holiday ☐   **Where?** ....
(b) Sickness ☐
(c) Injury ☐

2. DURING YOUR ABSENCE:

(a) did you **suffer** from:

Diarrhoea ☐   Vomiting ☐   Raised temperature ☐
Persistent cough ☐   Urinary infection ☐

Infections of: **Ears** ☐   **Nose** ☐   **Skin** ☐
                             Throat ☐   **Eyes** ☐
                             Boils ☐   Infected wounds ☐

(b) were you **in contact** with anyone (family, friends, etc.) suffering from:

Diarrhoea ☐   Vomiting ☐
Gastro-enteritis ☐   Food poisoning ☐

I have answered each question to the best of my knowledge and belief.

Signed: ....

Full name: ....

Date: ....

Occupational health: (d) Fitness certificate

Mr/Mrs/Ms: ....

Clock: .... employed as ....

has had a Health/Medical Examination and you are advised that he/she is:

- **FIT**
- **FIT SUBJECT TO RESTRICTIONS**
- **UNFIT**

for normal work.

Notes ....

....

....

....

....

....

.... Occupational Health Nurse

Medical Officer

.... Date

Prohibition notice

HEALTH AND SAFETY EXECUTIVE   Serial No. P

Health and Safety at Work etc. Act 1974, Sections 22–24

PROHIBITION NOTICE

Name and
address (See
Section 46)

(a) Delete as
    necessary
(b) Inspector's
    full name
(c) Inspector's
    official
    designation
(d) Official
    address

To ....
....
....
(a) Trading as ....
(b) ....
one of (c) ....
of (d) ....
.... Tel. no. ....

hereby give you notice that I am of the opinion that the
following activities,
namely: ....
....
....
....
which are (a) being carried on by you/about to be carried on by
you/under your control

(e) Location of
    activity

at (e) ....

Involve, or will involve (a) a risk/an imminent risk, of serious
personal injury. I am further of the opinion that the said matters
involve contraventions of the following statutory provisions:
....
....
....
....
because ....
....
....
and I hereby direct that the said activities shall not be carried on
by you or under your control (a) Immediately/after

(f) Date

(f) ....
unless the said contraventions and matters included in the
schedule, which forms part of this notice, have been remedied.

Signature: .... Date: ....

being an inspector appointed by an instrument in writing made
pursuant to Section 19 of the said Act and entitled to issue
LP2   this notice.

Reporting of Injuries, Diseases and Dangerous Occurrences Regulations 1995: (a) Report of an injury or dangerous occurrence

Health and Safety at Work etc. Act 1974
The Reporting of Injuries, Diseases and Dangerous Occurrences Regulations 1995

Report of an injury or dangerous occurrence

Filling in this form
This form must be filled in by an employer or other responsible person.

Part A

About you

1 What is your full name?

2 What is your job title?

3 What is your telephone number?

About your organisation
4 What is the name of your organisation?

5 What is its address and postcode?

6 What type of work does the organisation do?

Part B

About the incident

1 On what date did the incident happen?

/   /

2 At what time did the incident happen?
(Please use the 24-hour clock e.g. 0600)

3 Did the incident happen at the above address?
Yes ☐ Go to question 4
No ☐ Where did the incident happen?
   ☐ elsewhere in your organisation –
      give the name, address and
      postcode
   ☐ at someone else's premises – give
      the name, address and postcode
   ☐ in a public place – give details of
      where it happened

If you do not know the postcode, what is
the name of the local authority?

4 In which department, or where on the premises, did the incident happen?

Part C

About the injured person

If you are reporting a dangerous occurrence, go
to Part F.
If more than one person was injured in the same
incident, please attach the details asked for in
Part C and Part D for each injured person.

1 What is their full name?

2 What is their home address and postcode?

3 What is their home phone number?

4 How old are they?

5 Are they
   ☐ male?
   ☐ female?

6 What is their job title?

7 Was the injured person (tick only one box)
   ☐ one of your employees?
   ☐ on a training scheme? Give details:

   ☐ on work experience?
   ☐ employed by someone else? Give details of
      the employer:

   ☐ self-employed and at work?
   ☐ a member of the public?

Part D

About the injury
1 What was the injury (e.g. fracture, laceration)

2 What part of the body was injured?

3 Was the injury (tick the one box that applies)
   - ☐ a fatality?
   - ☐ a major injury or condition? (see accompanying notes)
   - ☐ an injury to an employee or self-employed person which prevented them doing their normal work for more than 3 days?
   - ☐ an injury to a member of the public which meant they had to be taken from the scene of the accident to a hospital for treatment?
4 Did the injured person (tick all the boxes that apply)
   - ☐ become unconscious?
   - ☐ need resuscitation?
   - ☐ remain in hospital for more than 24 hours?
   - ☐ none of the above.

Part E

About the kind of accident

Please tick the one box that best describes what happened, then go to Part G.
   - ☐ Contact with moving machinery or material being machined
   - ☐ Hit by a moving, flying or falling object
   - ☐ Hit by a moving vehicle
   - ☐ Hit something fixed or stationary

   - ☐ Injured while handling, lifting or carrying
   - ☐ Slipped, tripped or fell on the same level
   - ☐ Fell from a height
     How high was the fall?

     [   ] metres

   - ☐ Trapped by something collapsing

   - ☐ Drowned or asphyxiated
   - ☐ Exposed to, or in contact with, a harmful substance
   - ☐ Exposed to fire
   - ☐ Exposed to an explosion

   - ☐ Contact with electricity or an electrical discharge
   - ☐ Injured by an animal
   - ☐ Physically assaulted by a person

   - ☐ Another kind of accident (describe it in Part G)

Part F

Dangerous occurrences

Enter the number of the dangerous occurrence you are reporting. (The numbers are given in the Regulations and in the notes which accompany this form.)

[   ]

Part G

Describing what happened

Give as much detail as you can. For instance
- the name of any substance involved
- the name and type of any machine involved
- the events that led to the incident
- the part played by any people.

If it was a personal injury, give details of what the person was doing. Describe any action that has since been taken to prevent a similar incident. Use a separate piece of paper if you need to.

☐☐☐☐   [   ]

Part H

Your signature

Signature

[   ]

Date

[/   /   ]

Where to send the form

Please send it to the Enforcing Authority for the place where it happened. If you do not know the Enforcing Authority, send it to the nearest HSE office.

Reporting of Injuries, Diseases and Dangerous Occurrences Regulations 1995: (b) Report of a case of disease

Health and Safety at Work etc. Act 1974
The Reporting of Injuries, Diseases and Dangerous Occurrences Regulations 1995

Report of a case of disease

HSE
Health & Safety
Executive

Filling in this form
This form must be filled in by an employer or other responsible person.

Part A

About you

1 What is your full name?

2 What is your job title?

3 What is your telephone number?

About your organisation

4 What is the name of your organisation?

5 What is its address and postcode?

6 Does the affected person usually work at this address?
Yes ☐ Go to question 7
No ☐ where do they normally work?

7 What type of work does the organisation do?

Part B

About the affected person

1 What is their full name?

2 What is their date of birth?

☐ / ☐ / ☐

3 What is their job title?

☐☐☐

4 Are they
☐ male?
☐ female?

5 Is the affected person (tick one box)
☐ one of your employees?
☐ on a training scheme? Give details:

☐ on work experience?
☐ employed by someone else? Give details:

☐ other? Give details:

Part C

The disease you are reporting

1 Please give:
- the name of the disease, and the type of work it is associated with; or
- the name and number of the disease (from Schedule 3 of the Regulations - see the accompanying notes).

2 What is the date of the statement of the doctor who first diagnosed or confirmed the disease?

    /   /

3 What is the name and address of the doctor?

Part D

Describing the work that led to the disease

Please describe any work done by the affected person which might have led to them getting the disease.

If the disease is thought to have been caused by exposure to an agent at work (e.g. a specific chemical) please say what that agent is.

Give any other information which is relevant.

Give your description here

Continue your description here

Part E

Your signature

Signature

Date

    /   /

Where to send the form

Please send it to the Enforcing Authority for the place where the affected person works. If you do not know the Enforcing Authority, send it to the nearest HSE office.

For official use

Client number

Location number

Event number

☐ INV REP ☐ Y ☐ N

Part 4
Health and Safety Glossary

This glossary incorporates some of the more commonly used concepts and terms in occupational health and safety.

Absorption

The entry of a substance into the body. This may be by inhalation, pervasion (through the skin), ingestion, injection, inoculation and implantation.

☞ **1(c) Principal regulations**
   Chemicals (Hazard Information and Packaging for Supply) Regulations 2002
   Control of Asbestos at Work Regulations 2002
   Control of Lead at Work Regulations 2002
   Control of Substances Hazardous to Health Regulations 2002
   Ionising Radiations Regulations 1999

☞ **2(a) Health and safety in practice**
   Health records
   Health surveillance
   Local rules

☞ **2(b) Hazard checklists**
   Hazardous substances
   Radiation hazards

☞ **3(a) Tables and figures**
   Categories of danger – Chemicals (Hazard Information and Packaging for Supply) Regulations
   Hazardous substances that can be revealed by medical analysis

☞ **3(b) Forms**
   Control of Substances Hazardous to Health Regulations 2002 – Health Risk Assessment

☞ **4. Health and Safety Glossary**
   Acute effect
   Air sampling (air monitoring)
   Chemical hazards
   Dose

Dose–effect relationship
Dose–response relationship
Hazardous substances
Health risk assessment
Health surveillance
Route of entry
Substances hazardous to health
Target organs and target systems
Threshold dose
Toxicity
Toxicological assessment
Toxicology
Workplace exposure limit

Action levels, exposure action values and exposure limit values

These terms are commonly specified in regulations. Employers are required to take action wherever exposure of employees to an action level, action value or exposure action value is identified.

Action levels are specified with respect to exposure to asbestos and lead, thus:

Asbestos
The Control of Asbestos at Work Regulations 2002 specify one of the following cumulative exposures to asbestos over a continuous 12-week period when measured or calculated by a method approved by the HSC, namely:

(a) where the exposure is solely to chrysotile, 72 fibre-hours per millilitre of air;
(b) where the exposure is to any other form of asbestos either alone or in mixtures including mixtures of chrysotile with any other form of asbestos, 48 fibre-hours per millilitre of air; or

(c) where both types of exposure occur separately during the 12-week period concerned, a proportionate number of fibre-hours per millilitre of air.

 1(c) Principal regulations
Control of Asbestos at Work Regulations 2002

Lead
The Control of Lead at Work Regulations 2002 specify a blood lead concentration of:
(a) in respect of a woman of reproductive capacity, 25 μg/dl;
(b) in respect of a young person, 40 μg/dl;
(c) in respect of any other employee, 50 μg/dl.

 1(c) Principal regulations
Control of Lead at Work Regulations 2002

Noise
The Control of Noise at Work Regulations 2005 specify certain 'exposure action values' and 'exposure limit values' as follows:
- the lower exposure action values are:
  (a) a daily or weekly personal noise exposure of 80 dB (A-weighted); and
  (b) a peak sound pressure of 135 dB (C-weighted)
- the upper exposure action values are:
  (a) a daily or weekly personal noise exposure of 85 dB (A-weighted); and
  (b) a peak sound pressure of 137 dB (C-weighted)
- the exposure limit values are:
  (a) a daily or weekly personal noise exposure of 87 dB (A-weighted); and
  (b) a peak sound pressure of 140 dB (C-weighted).

Every employer must, when any of his employees is likely to be exposed to the first action level or above, or to the peak action level or above, ensure that a competent person makes a noise assessment.

☞ **1(c) Principal regulations**
   Control of Noise at Work Regulations 2005
☞ **2(b) Hazard checklists**
   Noise
☞ **3(a) Tables and figures**
   Decibels (addition of)
   Noise control methods
   Noise control programme (typical structure)
   Octave bands (standard range)
☞ **3(b) Forms**
   Noise exposure record

Vibration

The Control of Vibration at Work Regulations 2005 specify certain 'exposure limit values' and 'action values' thus:

For hand–arm vibration
   (a) the daily exposure limit value normalised to an 8-hour reference period is 5 m/s^2;
   (b) the daily exposure action value normalised to an 8-hour reference period is 2.5 m/s^2;
   (c) daily exposure shall be ascertained on the basis set out in Schedule 1 Part 1.

For whole body vibration
   (a) the daily exposure limit value normalised to an 8-hour reference period is 1.15 m/s^2;
   (b) the daily exposure action value normalised to an 8-hour reference period is 0.5 m/s^2;
   (c) daily exposure shall be ascertained on the basis set out in Schedule 2 Part 1.

☞ **1(c) Regulations**
   Control of Vibration at Work Regulations 2005

Active monitoring

A form of safety monitoring which entails a range of exercises directed at preventing accidents, including safety inspections, safety audits, safety tours and safety sampling exercises.

 2(a) Health and safety in practice
   Safety monitoring systems

Acute effect

A rapidly produced effect on the body following a single exposure to an offending or hazardous agent.

Aerosol

Any combination of particles carried in, or contained in, air. An aerosol may embrace liquid droplets as well as solid particles.

Air sampling (air monitoring)

The process of taking a sample of air for subsequent analysis. It may be undertaken on a short-term or long-term basis.

Short-term sampling (grab sampling, snap sampling) implies taking an immediate sample of air and, in most cases, passing it through a particular chemical reagent which responds to the contaminant being monitored.

Long-term sampling can be undertaken using personal sampling instruments or dosemeters, which are attached to the individual, and by the use of static sampling equipment located in the working area.

 1(c) Principal regulations
   Control of Asbestos at Work Regulations 2002
   Control of Lead at Work Regulations 2002

Control of Substances Hazardous to Health Regulations 2002
Dangerous Substances and Explosive Atmospheres Regulations 2002

 3(a) Tables and figures
Airborne contaminants: comparison of particle size ranges

Anthropometry

The study and measurement of body dimensions, the orderly treatment of the resulting data and the application of the data in the design of workspace layouts and equipment.

 3(a) Tables and figures
The total working system – areas of study
 4. Health and Safety Glossary
Ergonomics
Human factors

Atypical workers

A term used to describe workers and others who are not in normal daytime employment, together with shift workers, part-time workers and night workers.

[Working Time Regulations 1998]

Audiometry

The measurement of an individual's hearing acuity or ability over a range of frequencies. The determination of an individual's threshold levels for pure tones by air conduction under monoaural earphone listening conditions.

An audiogram, the outcome of an audiometric test, is used to assess the degree of hearing loss across the frequencies of interest, that is, the frequencies at which normal speech takes place, i.e. 0.5, 1 and 2 KHz. It is essentially a chart of a person's hearing threshold levels for pure tones of different frequencies.

 1(c) Principal regulations
    Control of Noise at Work Regulations 2005

 3(a) Tables and figures
    Decibels (addition of)
    Noise control programme – typical structure
    Octave bands (standard range)

 3(b) Forms
    Noise exposure record

 4. Health and Safety Glossary
    Health surveillance
    Noise-induced hearing loss (occupational deafness)
    Octave band analysis
    Primary monitoring
    Reduced time exposure (limitation)

Auditing

The structured process of collecting independent information on the efficiency, effectiveness and reliability of the total safety management system and drawing up plans for corrective action. [*Successful health and safety management (HS(G)65)*]

 2(a) Health and safety in practice
    BS 8800: Guide to Occupational Health and Safety Management Systems
    OHSAS 18001: A Pro-active Approach to Health and Safety Management
    Safety monitoring systems
    Successful health and safety management (HS(G)65)

 3(a) Tables and figures
    Key elements of successful health and safety management

Biological hazards

Ill-health can result from exposure to biological agents, such as bacteria, viruses and dusts. Biological hazards can be classified according to origin:

- animal-borne – e.g. anthrax, brucellosis
- human-borne – e.g. viral hepatitis
- vegetable-borne – e.g. aspergillosis (farmers' lung).

Biological monitoring

A regular measuring activity where selected validated indicators of the uptake of toxic substances are determined in order to prevent health impairment.

Biological monitoring may feature as part of the health surveillance procedures required under the Control of Substances Hazardous to Health (COSHH) Regulations.

It may be undertaken through the determination of the effects certain substances produce on biological samples of exposed individuals, and these determinations are used as biological indicators.

Biological samples where indicators may be determined consist of:

- blood, urine, saliva, sweat, faeces
- hair, nails, and
- expired air.

Indicators of internal dose can be divided into:

- true indicators of dose, i.e. capable of indicating the quantity of the substance at the sites of the body where it exerts its effect,
- indicators of exposure, which can provide an indirect estimate of the degree of exposure, since the levels of substances in the biological samples closely correlate with levels of environmental pollution, and
- indicators of accumulation that can provide an evaluation of the concentration of the substance in organs and/or

tissues from which the substance, once deposited, is slowly released.

It also includes the measuring of a person's blood-lead concentrations or urinary lead concentration in accordance in either case with the method known as atomic absorption spectrometry.

☞ **1(c) Principal regulations**
> *Control of Lead at Work Regulations 2002*
> *Control of Substances Hazardous to Health Regulations 2002*

☞ **2(a) Health and safety in practice**
> *Health records*
> *Health surveillance*

☞ **2(b) Hazard checklists**
> *Hazardous substances*
> *Hazardous substances that can be revealed by medical analysis*

☞ **3(b) Forms**
> *Control of Substances Hazardous to Health Regulations 2002 – Health Risk Assessment*

☞ **4. Health and Safety Glossary**
> *Absorption*
> *Biological hazards*
> *Dose*
> *Health risk assessment*
> *Health surveillance*
> *Occupational health*
> *Route of entry*
> *Substances hazardous to health*

British Standards

The British Standards Institution produces safety standards and codes through committees formed to deal with a specific matter or subject, such as machinery safety.

Standards contain details relating to, for instance, the construction of, and materials incorporated in, an item and, where necessary, prescribe methods of testing to establish compliance.

Codes deal with safe working practices and systems of work.

British Standards and Codes have no legal status, but can be interpreted by the courts as being the authoritative guidance on a particular matter.

Carcinogen

This means:
    (a) any substance or preparation which if classified in accordance with the classification provided for by Regulation 5 of the Chemicals (Hazard Information and Packaging for Supply) Regulations 1994 would be in the category of danger, carcinogenic (category 1) or carcinogenic (category 2) whether or not the substance or preparation would be required to be classified under those Regulations; or
    (b) any substance or preparation:
      (i) listed in Schedule 1; and
      (ii) arising from a process specified in Schedule 1 which is a substance hazardous to health.
[Control of Substances Hazardous to Health Regulations 2002]

☞ **1(c) Principal regulations**
    Chemicals (Hazard Information and Packaging for Supply) Regulations 2002
    Control of Substances Hazardous to Health Regulations 2002
☞ **2(b) Hazard checklists**
    Hazardous substances
☞ **3(a) Tables and figures**
    Categories of danger – Chemicals (Hazard Information and Packaging for Supply) Regulations 2002
    Hazardous substances that can be revealed by medical analysis
    Local exhaust ventilation systems
☞ **3(b) Forms**
    Control of Substances Hazardous to Health Regulations 2002 – Health Risk Assessment

4. Health and Safety Glossary

Acute effect
Air sampling (air monitoring)
Chemical hazards
Chronic effect
Containment
Dose
Dose–response relationship
Hazardous substances
Health risk assessment
Health surveillance
Local exhaust ventilation (LEV)
Occupational health
Prohibition
Route of entry
Target organs and target systems
Toxicity
Toxicological assessment

CE marking

A specific form of marking which must be applied to a wide range of equipment, such as electrical equipment, indicating that the equipment complies with all the requirements of regulations, e.g. the Electrical Equipment (Safety) Regulations 1994, which implement that particular European Council Directive.

Change of process

A common strategy in protecting both the health and safety of people exposed to hazards. In this case, improved design or process engineering can result in changes to provide better protection, as in the case of dusty processes or those producing noise.

1(c) Principal regulations

Control of Asbestos at Work Regulations 2002
Control of Lead at Work Regulations 2002
Control of Noise at Work Regulations 2005
Control of Substances Hazardous to Health Regulations 2002
Control of Vibration at Work Regulations 2005
Ionising Radiations Regulations 1999
Manual Handling Operations Regulations 1992

Chemical hazards

These are hazards arising from the use and storage of chemical substances and which result in a range of chemical poisonings and other forms of disease or condition. Exposure to these hazards may result in dermatitis, occupational cancers and respiratory disorders.

 1(c) Principal regulations

Chemicals (Hazard Information and Packaging for Supply) Regulations 2002
Control of Lead at Work Regulations 2002
Control of Substances Hazardous to Health Regulations 2002
Dangerous Substances and Explosive Atmospheres Regulations 2002

 2(b) Hazard checklists

Hazardous substances

 3(a) Tables and figures

Airborne contaminants: comparison of particle size ranges
Categories of danger – Chemicals (Hazard Information and Packaging for Supply) Regulations 2002
Hazardous substances that can be revealed by medical analysis
Safety data sheets – obligatory headings
Safety signs

3(b) Forms
Control of Substances Hazardous to Health Regulations 2002 – Health Risk Assessment

 4. Health and Safety Glossary
Absorption
Acute effect
Aerosol
Air sampling (air monitoring)
Carcinogen
Change of process
Chronic effect
Dilution ventilation
Dose
Dose–effect relationship
Dose–response relationship
Elimination
Hazardous substances
Health risk assessment
Health surveillance
Local effect
Local exhaust ventilation (LEV)
Neutralisation
Occupational health
Occupational hygiene
Prescribed disease
Primary monitoring
Prohibition
Reportable disease
Route of entry
Secondary monitoring
Substances hazardous to health
Substitution
Target organs and target systems
Threshold dose
Toxicological assessment
Toxicity
Workplace exposure limit

Chronic effect

An effect on the body as a result of prolonged exposure or repeated exposure of long duration.

Comfort

A subjective assessment of the conditions in which a person works, sleeps, relaxes, travels, etc. and which varies according to age, state of health and vitality. Comfort is directly related to environmental factors such as temperature, ventilation and humidity.

 1(c) Principal regulations
   Building Regulations 2000
   Workplace (Health, Safety and Welfare) Regulations 1992

Comfort ventilation

The process of providing sufficient air for people to breathe and, to some extent, regulating temperature. It is directly related to the number of air changes per hour in a workplace according to the external ambient air temperature and the actual rate of air movement. Rates of air change will, in most cases, vary from summer to winter in order to maintain comfort.

 1(c) Principal regulations
   Building Regulations 2000
   Workplace (Health, Safety and Welfare) Regulations 1992
 3(a) Tables and figures
   Air changes per hour (comfort ventilation)
   Maximum permissible wet bulb globe temperatures
   Optimum working temperatures
 4. Health and Safety Glossary
   Comfort

Compartmentation

A structural process, designed to limit the spread of fire within a building, which divides the building into fire-resistant cells or units, both vertically and horizontally. It is further used to segregate high risk areas of a building from other areas.

 1(c) Principal regulations
   Building Regulations 2000
   Regulatory Reform (Fire Safety) Order 2005
 2(b) Hazard checklists
   Fire safety
   Flammable substances

Confined space

This is defined as 'a place which is substantially, though not always entirely, enclosed, and where there is a risk that anyone who may enter the space could be injured due to fire or explosion, overcome by gas, fumes, vapour, or the lack of oxygen, drowned, buried under free-flowing solids, such as grain, or overcome due to high temperature'.

Under the Confined Spaces Regulations 1997 employers must:
   (a) avoid employees from entering confined spaces, for example, by undertaking the work from outside;
   (b) follow a safe system of work, e.g. a Permit to Work system, if entry to a confined space is unavoidable; and
   (c) put in place adequate emergency arrangements before work starts, which will also safeguard any rescuers.
The regulations are accompanied by an ACOP and HSE Guidance.

 1(c) Principal regulations
   Confined Spaces Regulations 1997
 4. Health and Safety Glossary
   Permit to work

Construction work

The carrying out of any building, civil engineering or engineering construction work, including any of the following:

  (a) the construction, alteration, conversion, fitting out, commissioning renovation, repair, upkeep, redecoration or other maintenance (including cleaning) which involves the use of water or an abrasive at high pressure or the use of substances classified as corrosive or toxic for the purposes of Regulation 7 of the Chemicals (Hazard Information and Packaging for Supply) Regulations 1994, decommissioning, demolition or dismantling of a structure;

  (b) the preparation for an intended structure, including site clearance, exploration, investigation (but not site survey) and excavation, and laying or installing the foundations of a structure;

  (c) the assembly of prefabricated elements to form a structure or the disassembly of prefabricated elements which, immediately before such disassembly, formed a structure;

  (d) the removal of a structure or part of a structure or of any product or waste resulting from demolition or dismantling of a structure or from disassembly of prefabricated elements which, immediately before such disassembly, formed a structure;

  (e) the installation, commissioning, maintenance, repair or removal of mechanical, electrical, gas, compressed air, hydraulic, telecommunications, computer or similar services which are normally fixed within or to a structure.

[Construction (Design and Management) Regulations 1994]

 1(c) Principal regulations
   Building Regulations 2000
   Chemicals (Hazard Information and Packaging for Supply) Regulations 2002
   Construction (Design and Management) Regulations 1994

Construction (Health, Safety and Welfare) Regulations 1996
Work at Height Regulations 2005

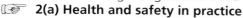 **2(a) Health and safety in practice**
Competent persons
Health and safety file
Health and safety plans
Method statements

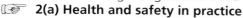 **2(b) Hazard checklists**
Construction activities
Maintenance work

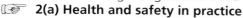 **3(a) Tables and figures**
Construction (Design and Management) Regulations 1994 –
How to decide when the exceptions to the CDM Regulations apply
Places of work requiring inspection by a competent person
under Regulation 29(1) of the Construction (Health, Safety
and Welfare) Regulations 1996

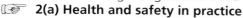 **3(b) Forms**
Construction (Design and Management) Regulations 1994 –
Notification of Project (Form 10)

Contact hazard

A hazard arising from contact with a machine at a particular
point arising from sharp surfaces, sharp projections, heat and
extreme cold.

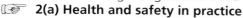 **1(c) Principal regulations**
Provision and Use of Work Equipment Regulations 1998

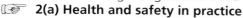 **4. Health and Safety Glossary**
Machinery hazards
Machinery guards
Machinery safety devices
Non-mechanical hazards
Non-operational parts (machinery)
Operational parts (machinery)

Containment/enclosure

The structural prevention of fire spread to other parts of a premises through the use of fire doors, fireproof compartments and other fire-resistant structures.

Total containment or enclosure of a dangerous processing operation is effected by the use of bunds, bulk tanks and pipework to deliver a liquid directly into a closed production vessel.

 1(c) Principal regulations
Building Regulations 2000
Regulatory Reform (Fire Safety) Order 2005

Danger

Liability of exposure to harm; a thing that causes peril.

When applied to machinery in motion is a situation in which there is a reasonably foreseeable risk of injury from mechanical hazards associated with contact with it or being trapped between the machinery and any material in or at the machinery, or any fixed structure. Or being struck by, or entangled in or by any material in motion in the machinery or being struck by parts of the machinery ejected from it, or being struck by material ejected from the machinery.

[BS EN ISO 12100: Safety of machinery]

Dangerous occurrence

An event listed in Schedule 2 of the Reporting of Injuries, Diseases and Dangerous Occurrences Regulations (RIDDOR) 1995. It is an event with particularly significant potential for death and/or major injury, such as the collapse or overturning of lifting machinery, unintentional explosions, gassing accidents and boiler explosions.

Under RIDDOR, dangerous occurrences are classified in five groups – general, those relating to mines, those relating to quarries, those relating to relevant transport systems and those is respect of offshore workplaces.

☞ **1(c) Principal regulations**
> *Reporting of Injuries, Diseases and Dangerous Occurrences Regulations 1995*

☞ **3(a) Tables and figures**
> *Reporting of Injuries, Diseases and Dangerous Occurrences Regulations – Reporting requirements*

☞ **3(b) Forms**
> *Reporting of Injuries, Diseases and Dangerous Occurrences Regulations 1995*
> *(a) Report of an Injury or Dangerous Occurrence (Form 2508)*

Dilution ventilation

In certain situations it may not be possible to use a local exhaust ventilation system to remove airborne contaminants. Where the quantity of contaminant is small, uniformly evolved and of low toxicity, it may be possible to dilute the contaminant by inducing large volumes of air to flow through the contaminated region.

Dilution ventilation is most successfully used to control vapours from low toxicity solvents, but is seldom successfully applied to dust and fumes.

Display screen equipment

Any alphanumeric or graphic display screen, regardless of the display process involved.

[Health and Safety (Display Screen Equipment) Regulations 1992]

The regulations apply only to 'users' and 'operators' of display screen equipment:

- a 'user' means an employee who habitually uses display screen equipment as a significant part of his normal work
- an 'operator' means a self-employed person who habitually uses display screen equipment as a significant part of his normal work.

☞ **1(c) Principal regulations**
   Health and Safety (Display Screen Equipment) Regulations 1992

☞ **2(c) Hazard checklists**
   Display screen equipment

☞ **3(a) Tables and figures**
   Health and Safety (Display Screen Equipment) 1992:
   (a) *Display screen equipment workstation – design and layout*
   (b) *Seating and posture for typical office tasks*

Dose

The level of environmental contamination or offending agent related to the duration of exposure to same.

Dose = Level of environmental contamination
       $\times$ Duration of exposure

The term is used in the case of physical stressors e.g. noise, chemical stressors, e.g. gases and biological stressors, e.g. bacteria.

☞ **1(c) Principal regulations**
   Control of Asbestos at Work Regulations 2002
   Control of Lead at Work Regulations 2002
   Control of Noise at Work Regulations 2005
   Control of Substances Hazardous to Health Regulations 2002
   Control of Vibration at Work Regulations 2005
   Ionising Radiations Regulations 1999

2(a) Health and safety in practice
Dose Record
Health records
Health surveillance
Local rules
Risk assessment

 2(c) Hazard checklists
Hazardous substances
Noise
Radiation hazards

 3(a) Tables and figures
Hazardous substances that can be revealed by medical analysis

 3(b) Forms
Control of Substances Hazardous to Health Regulations 2002 – Health Risk Assessment
Noise exposure record

 4. Health and Safety Glossary
Absorption
Action levels, exposure action levels and exposure limit values
Acute effect
Air sampling (air monitoring)
Audiometry
Biological hazards
Biological monitoring
Carcinogen
Chemical hazards
Chronic effect
Dose–effect relationship
Dose–response relationship
Hazardous substances
Health risk assessment
Health surveillance
Ionising radiation
Noise-induced hearing loss (occupational deafness)
Occupational hygiene
Primary monitoring
Reduced time exposure (limitation)
Route of entry

Secondary monitoring
Segregation
Substances hazardous to health
Substitution
Target organs and target systems
Threshold dose
Toxicity
Toxicological assessment
Toxicology

Dose–effect relationship

Estimation of the relationship between the specific dose of a contaminant and its effects on the human body is based on the degree of association existing, firstly, between an indicator of dose, i.e. urine, faeces, blood, saliva and, secondly, an indicator of effect on the body, e.g. respiratory difficulties, unconsciousness, headaches.

The study of this relationship will show the particular concentration of a toxic substance at which the indicator of effect exceeds the value currently accepted as 'normal'.

See cross references for *Dose*

Dose–response relationship

Consideration of threshold limits of exposure or dose, which most people can tolerate without either short-term or long-term damage to their health, is a basic feature of the prevention and control of occupational diseases.

For many chemicals commonly used, it is possible to establish a relationship or link between the dose received and the body's response (e.g. coughing, lachrymation), a characteristic known as the 'dose–response relationship'.

Where dose is plotted against response in a graphical form, with many dusts, for instance, the response is directly proportion to the dose. In the case of other environmental contaminants, the dose response curve remains at a level of no response at a point greater than zero on the dose axis. This point of cut-off identifies the threshold dose. After reaching the threshold dose, the body's response rises dramatically.

See cross references for *Dose*

Elimination

A prevention strategy in the use of hazardous substances whereby substances no longer in use, or which can be replaced by less hazardous substances, are eliminated from an organisation's inventory and stock of substances.

 2(b) Hazard checklists
   Hazardous substances
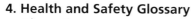 **4. Health and Safety Glossary**
   Absorption
   Hazardous substances
   Health risk assessment

Emergency lighting

This form of lighting is provided to ensure safety when a normal lighting installation fails.

Standby lighting enables essential work to continue, the illuminance required depending upon the nature of the work. It may be between 5% and 100% of the illuminance provided by the normal lighting installation.

Escape lighting enables a building to be evacuated safely, and may take the form of battery or generator-powered installations.

1(b) Principal regulations
Building Regulations 2000
Regulatory Reform (Fire Safety) Order 2005
Workplace (Health, Safety and Welfare) Regulations 1992

Ergonomics

Ergonomics can be defined in several ways:
- the scientific study of work
- human factors engineering
- the study of the man–machine interface
- the scientific study of the interrelationships between people and their work.

Ergonomics takes into account:
- the human system
- environmental factors
- the man–machine interface, and
- the total working system.

 1(c) Principal regulations
Health and Safety (Display Screen Equipment) Regulations 1992
Management of Health and Safety at Work Regulations 1999

 3(a) Tables and figures
Health and Safety (Display Screen Equipment) Regulations 1992
(a) Display screen equipment workstation – design and layout
(b) Seating and posture for typical office tasks
The Total Working System – Areas of study

4. Health and Safety Glossary
Anthropometry
Human factors
Job design

Fail safe

A design feature of machinery whereby any failure in, or interruption of, the power supply to a safeguard will result in the prompt stopping or, where appropriate, stopping and reversal of the movement of the dangerous parts before injury can occur, or the safeguard remaining in position to prevent access to the danger point or area.

 1(c) Principal regulations
   Provision and Use of Work Equipment Regulations 1998
 2(b) Hazard checklists
   Work equipment
 4. Health and Safety Glossary
   Intrinsic safety
   Machinery hazards
   Machinery guards
   Machinery safety devices

Fire instructions

A notice informing people of the action they should take on either hearing a fire alarm or discovering a fire.

 3(a) Tables and figures
   Fire instruction notice

First aid

The skilled application of accepted principles of treatment on the occurrence of an accident or in the case of sudden illness, using facilities and materials available at the time.

The principal aims of first aid are:
- to sustain life
- to prevent deterioration in an existing condition, and
- to promote recovery.

Under the Health and Safety (First Aid) Regulations 1981, first aid means:

- in cases where a person will need help from a medical practitioner or nurse, treatment for the purpose of preserving life and minimising the consequences of injury or illness until such help is obtained, and
- treatment of minor injuries which would otherwise receive no treatment or which do not need treatment by a medical practitioner or nurse.

 1(c) Principal regulations
Health and Safety (First Aid) Regulations 1981

Fracture mechanics

A branch of engineering science concerned with the study of material failures and factors which determine the probability of catastrophic failure of various structural components. The results of these studies can be used in the design of structures, machinery and lifting appliances.

Gas incident

Any death or any major injury which has arisen out of or in connection with the gas distributed, filled, imported or supplied, as the case may be, by a conveyor of flammable gas through a fixed pipe distribution system, or a filler, importer or supplier (other than by means of retail trade) of a refillable container containing liquefied petroleum gas.

[Reporting of Injuries, Diseases and Dangerous Occurrences Regulations 1995]

 1(c) Principal regulations
Reporting of Injuries, Diseases and Dangerous Occurrences Regulations 1995

Generic risk assessment

A risk assessment produced once only for a given activity, type of workplace or specific work group. This is particularly appropriate where organisations run a range of similar workplaces in different locations, e.g. maintenance workshops, undertake activities which are standard, e.g. tyre fitting and employ people to carry out the same type of work in different locations, e.g. postmen.

For generic assessments to be effective:
    (a) 'worst case' situations must be considered; and
    (b) provision should be made within the generic risk assessment to monitor implementation of the recommended preventive measures and controls which are relevant to a particular workplace, work activity or work group.

 1(c) Principal regulations
    Management of Health and Safety at Work Regulations 1999
 2(a) Health and safety in practice
    BS 8800: Guide to Occupational Health and Safety Management Systems
    OHSAS 18001: A Pro-Active Approach to Health and Safety Management
    Risk assessment
    Successful Health and Safety Management
 4. Health and Safety Glossary
    Atypical workers
    Hazard
    Risk

Hazard

Something with the potential to cause harm. This can include substances or machines, methods of work and other aspects of work organisation.

The result of a departure from the normal situation, which has the potential to cause death, injury, damage or loss.

The physico-chemical or chemical property of a dangerous substance which has the potential to give rise to fire, explosion, or other events which can result in harmful physical effects of a kind similar to those which can be caused by fire or explosion, affecting the safety of a person.

[Dangerous Substances and Explosive Atmospheres Regulations 2002]

In relation to a substance, means the intrinsic property of that substance which has the potential to cause harm to the health of a person.

[Control of Substances Hazardous to Health Regulations 2002]

☞ **1(c) Principal regulations**
Control of Major Accident Hazards Regulations 1999
Control of Substances Hazardous to Health Regulations 2002
Dangerous Substances and Explosive Atmospheres Regulations 2002
Management of Health and Safety at Work Regulations 1999

☞ **2(a) Health and safety in practice**
Risk assessment
Safe systems of work
Safety monitoring systems

☞ **4. Health and Safety Glossary**
Risk

Hazardous substances

Hazardous substances and preparations are classified according to their category of danger under the Chemicals (Hazard Information and Packaging for Supply) (CHIP) Regulations 1994

Classification is on the basis of
• physico-chemical properties – explosive, oxidising, extremely flammable, highly flammable, flammable

- health effects – very toxic, toxic, harmful, corrosive, irritant, sensitising, carcinogenic, mutagenic, toxic for reproduction
- dangerous for the environment.

Certain substances may have a double classification, e.g. 'toxic' and 'flammable'.

☞ 1(c) Principal regulations

Chemicals (Hazard Information and Packaging for Supply) Regulations 2002

Control of Substances Hazardous to Health Regulations 2002

Dangerous Substances and Explosive Atmospheres Regulations 2002

Highly Flammable Liquids and Liquefied Petroleum Gases Regulations 1972

☞ 2(b) Hazard checklists

Flammable substances

Hazardous substances

☞ 3(a) Tables and figures

Airborne contaminants: comparison of particle size ranges

Categories of danger – Chemicals (Hazard Information and Packaging for Supply) Regulations 2002

Hazardous substances that can be revealed by medical analysis

Safety data sheets – obligatory headings

☞ 3(b) Forms

Control of Substances Hazardous to Health Regulations 2002 – Health Risk Assessment

☞ 4. Health and Safety Glossary

Absorption

Acute effect

Aerosol

Air sampling (air monitoring)

Chemical hazards

Chronic effect

Dose

Dose–effect relationship

Dose–response relationship

Elimination

Health risk assessment
Health surveillance
Local effect
Long-term exposure limit
Occupational hygiene
Primary monitoring
Route of entry
Secondary monitoring
Substances hazardous to health
Substitution
Target organs and target systems
Threshold dose
Toxicological assessment
Toxicology
Workplace exposure limits

Health risk assessment

Where there may be a risk of exposure of employees to a substance hazardous to health an employer must make a suitable and sufficient assessment of the risks created by that work to the health of those employees and the steps that need to be taken to meet the requirements of the regulations.

An assessment of the risks created by any work should involve:
   (a) consideration of:
      (i) which substances or types of substance (including biological agents) employees are liable to be exposed to (taking into account the consequences of possible failure of any control measure provided to meet the requirements of Regulation 7);
      (ii) what effects those substances can have on the body;
      (iii) where the substances are likely to be present and in what form;
      (iv) the ways in which and the extent to which any groups of employees or other persons could potentially be exposed, taking into account the nature of

the work and process, and any reasonably foreseeable deterioration in, or failure of, any control measure provided for the purposes of Regulation 7;

(b) an estimate of exposure, taking into account engineering measures and systems of work currently employed for controlling potential exposure;

(c) where valid standards exist, representing adequate control, comparison of the estimate with those standards.

[Control of Substances Hazardous to Health Regulations 2002 and ACOP]

☞ **1(c) Principal regulations**
Control of Asbestos at Work Regulations 2002
Control of Lead at Work Regulations 2002
Control of Substances Hazardous to Health Regulations 2002
Management of Health and Safety at Work Regulations 1999

☞ **2(a) Health and safety in practice**
Health records
Health surveillance
Risk assessment

☞ **2(b) Hazard checklists**
Hazardous substances

☞ **3(a) Tables and figures**
Airborne contaminants: comparison of particle size ranges
Categories of danger: Chemicals (Hazard Information and Packaging for Supply) Regulations
Hazardous substances that can be revealed by medical analysis
Safety data sheets – obligatory headings

☞ **3(b) Forms**
Control of Substances Hazardous to Health Regulations 2002 – Health Risk Assessment

☞ **4. Health and Safety Glossary**
Biological monitoring
Chemical hazards
Dose

Dose–response relationship
Hazardous substances
Health surveillance
Primary monitoring
Route of entry
Secondary monitoring
Substances hazardous to health
Threshold dose
Toxicity
Toxicological assessment
Workplace exposure limit

Health surveillance

The specific health examination at a predetermined frequency of those at risk of developing further ill health or disability, e.g. employees exposed to chemical hazards, and those actually or potentially at risk by virtue of the type of work they undertake during their employment, e.g. radiation workers.

The assessment of the state of health of an employee, as related to exposure to substances hazardous to health, and includes biological monitoring.

[Control of Substances Hazardous to Health Regulations 2002]

 1(c) Principal regulations
Control of Asbestos at Work Regulations 2002
Control of Lead at Work Regulations 2002
Control of Noise at Work Regulations 2005
Control of Substances Hazardous to Health Regulations 2002
Control of Vibration at Work Regulations 2005
Ionising Radiations Regulations 1999
Management of Health and Safety at Work Regulations 1999
 2(a) Health and safety in practice
Dose Record
Health records
Health surveillance

 **3(a) Tables and figures**
    Hazardous substances that can be revealed by medical analysis
3(b) Forms
    Occupational health:
    (a) Pre-employment health questionnaire
    (b) Health questionnaire
    (c) Food handler's clearance certificate
    (d) Fitness certificate
4. Health and Safety Glossary
    Active monitoring
    Audiometry
    Biological monitoring
    Dose–response relationship
    Hazardous substances
    Health risk assessment
    Local effect
    Occupational health
    Prescribed disease
    Primary monitoring
    Reportable disease
    Secondary monitoring
    Toxicological assessment

Hot work

The use of a range of equipment which may produce direct flames, heat, sparks, and arcing, and involving processes such as welding, cutting, brazing, soldering and the boiling of bitumen.

2(a) Health and safety in practice
    Method statements
    Risk assessment
    Safe systems of work
    Safety monitoring systems
2(b) Hazard checklists
    Fire safety
    Flammable substances

4. Health and Safety Glossary
Permit to work
Personal protective equipment

Human factors

A term used to cover a range of issues including:
- the perceptual, physical and mental capabilities of people and the interaction of individuals with their job and working environment
- the influence of equipment and system design on human performance, and
- the organisational characteristics which influence safety-related behaviour.

These are affected by:
- the system for communication within the organisation, and
- the training systems and procedures in operation

all of which are directed at preventing human error.

 1(c) Principal regulations
Health and Safety (Consultation with Employees) Regulations 1996
Health and Safety (Display Screen Equipment) Regulations 1992
Health and Safety (Information for Employees) Regulations 1998
Management of Health and Safety at Work Regulations 1999
Safety Representatives and Safety Committees Regulations 1977

 2(a) Health and safety in practice
Accident investigation procedures
Consequence analysis
Health and safety training
Information and instruction
Joint consultation

Management oversight and risk tree analysis
OHSAS 18001: A Pro-active Approach to Health and Safety
*   Management*
Risk assessment
Risk management
Safe systems of work
Successful health and safety management
Technique of human error rate probability
Total loss control

☞ **3(a) Tables and figures**
Health and Safety (Display Screen Equipment) Regulations
*   1992*
(a) Display screen equipment workstation – design and layout
(b) Seating and posture for typical office tasks
The Total working system – Areas of study

☞ **3(b) Forms**
Job safety analysis record

☞ **4. Health and Safety Glossary**
Anthropometry
Atypical workers
Ergonomics
Job design
Job safety analysis
Job safety instructions
Joint consultation
Organisational characteristics
Safety culture
Safety propaganda

Illuminance

The quantity of light flowing from a source, such as a light
bulb. It is sometimes referred to as 'luminous flux' or light flow,
and measured in lux.

☞ **1(c) Principal regulations**
Workplace (Health, Safety and Welfare) Regulations 1992

 **3(a) Tables and figures**
   Average illuminances and minimum measured illuminances
   Maximum ratios of illuminance for adjacent areas
 **4. Health and Safety Glossary**
   Emergency lighting

Impulse noise

Noise which is produced by widely spaced impacts between, for instance, metal parts, such as drop hammers.

 **1(c) Principal regulations**
   Control of Noise at Work Regulations 2005
 **2(b) Hazard checklists**
   Noise
 **3(a) Tables and figures**
   Decibels (addition of)
   Noise control methods
   Noise control programme – Typical structure
   Octave bands (standard range)
 **3(b) Forms**
   Noise Exposure Record
 **4. Health and Safety Glossary**
   Action levels, exposure action levels and exposure limit values
   Audiometry
   Dose
   Noise-induced hearing loss (occupational deafness)
   Octave band analysis

Incident

- An event which does not result in injury, damage or loss but which may cause interruption of the work process.
- An undesired event that could, or does, result in loss.

- An undesired event that could, or does, downgrade the efficiency of the business operation.

 2(a) Health and safety in practice
Accident investigation procedures
Consequence analysis
Event tree analysis
Failure mode and effect analysis
Fault tree analysis
Major incidents
Management oversight and risk tree analysis
Safe systems of work
Total loss control

 3(a) Tables and figures
Accident indices
Accident ratios

 4. Health and Safety Glossary
Near miss
Reportable event
Risk avoidance

Inspection (work equipment)

In relation to an inspection under Regulation 6:
- means such visual or more rigorous inspection by a competent person as is appropriate for the purpose described in that paragraph;
- where it is appropriate to carry out testing for the purpose, includes testing the nature and extent of which are appropriate for the purpose.

[Provision and Use of Work Equipment Regulations 1998]

 1(c) Principal regulations
Lifting Operations and Lifting Equipment Regulations 1998
Lifts Regulations 1997
Pressure Systems Safety Regulations 2000

Provision and Use of Work Equipment Regulations 1998
Simple Pressure Vessels (Safety) Regulations 1991
Work at Height Regulations 2005

 2(a) Health and safety in practice
Planned preventive maintenance
 2(b) Hazard checklists
Display screen equipment
Electrical equipment
Mobile mechanical handling equipment (lift trucks, etc)
Work equipment

Intrinsic safety

A concept based on the principle that sparks whose electrical parameters (voltage, current, energy) do not exceed certain level are incapable of igniting a flammable atmosphere. It is applied as a concept to low energy circuits, such as instrumentation and control systems.

Intrinsically safe equipment

The use of electrical equipment in flammable atmospheres requires that such equipment should be intrinsically safe, i.e. not provide a source of ignition. On this basis, such equipment must be flameproofed and intrinsically safe for use in potentially flammable hazardous areas. These areas are classified according to a graded possibility of an explosive gas or vapour concentration occurring.

 1(c) Principal regulations
Dangerous Substances and Explosive Atmospheres Regulations 2002
Electricity at Work Regulations 1989
 2(b) Hazard checklists
Electrical equipment

Ionising radiation

The transfer of energy in the form of particles or electromagnetic waves of a wavelength of 100 nanometres or less or a frequency of 3×10^{15} hertz or more capable of producing ions directly or indirectly.

[Ionising Radiations Regulations 1999]

 1(c) Principal regulations
Ionising Radiations Regulations 1999
 2(b) Hazard checklists
Radiation hazards
 3(a) Tables
Electromagnetic spectrum

Isolation

A commonly used control measure against identified risks implying the isolation or segregation of people from the particular hazard by, for instance, the use of remote control handling systems, enclosure of a plant or process producing harmful substances, the installation of high risk processing plants in remote parts of a country and enclosure of an individual in an acoustic booth or enclosure to protect against noise exposure.

Electrical isolation implies the disconnection and separation of an electrical appliance from every source of electrical energy in such a way that both disconnection and separation are secure.

 1(c) Principal regulations
Control of Noise at Work Regulations 2005
Electricity at Work Regulations 1989
Ionising Radiations Regulations 1999
 4. Health and Safety Glossary
Risk avoidance
Segregation

Job design

In the design of jobs, the following major considerations should be made:

- identification and comprehensive analysis of the critical tasks expected of individuals and appraisal of likely errors;
- evaluation of required operator decision making and the optimum balance between human and automatic contributions to safety actions;
- application of ergonomic principles to the design of man–machine interfaces, including displays of plant process information, control devices and panel layouts;
- design and presentation of procedures and operating instructions;
- organisation and control of the working environment, including the extent of the workspace, access for maintenance work and the effects of noise, lighting and thermal conditions;
- provision of the correct tools and equipment;
- scheduling of work patterns, including shift organisation, control of fatigue and stress, and arrangements for emergency operations/situations;
- efficient communications, both immediate and over periods of time.

[Reducing error and influencing behaviour – HS(G)48]

☞ **1(c) Principal regulations**
   Management of Health and Safety at Work Regulations 1999
☞ **1(d) Approved codes of practice**
   Management of health and safety at work
☞ **1(e) HSE guidance notes**
   Reducing error and influencing behaviour
☞ **4. Health and Safety Glossary**
   Anthropometry
   Ergonomics
   Human factors
   Job safety analysis
   Job safety instructions

Job safety analysis

A technique in the design of safe systems of work which identifies all the accident prevention measures appropriate to a particular job or area of work activity and the behavioural factors which most significantly influence whether or not these measures are taken.

 2(a) Health and safety in practice
   Safe systems of work
   Safety monitoring
 3(c) Forms
   Job safety analysis record
 4. Health and Safety Glossary
   Job design
   Job safety instructions

Job safety instructions

Job safety instructions are commonly one of the outcomes of job safety analysis, a technique used in the design of safe systems of work. Such instructions inform operators of specific risks at different stages of a job and advise of the precautions necessary to be taken at each stage.

Job safety instructions should be imparted to operators at the induction stage of their health and safety training and regularly reinforced.

 2(a) Health and safety in practice
   Safe systems of work

Joint consultation

An important means of improving motivation of people by enabling them to participate in planning work and setting objectives.

The processing of consulting with employees and others on health and safety procedures and systems. This may take place through consultation by an employer with trade-union-appointed safety representatives, non-trade-union representatives of employee safety and through the operation of a health and safety committee.

Legal and practical requirements relating to joint consultation are laid down in the Safety Representatives and Safety Committees Regulations 1977 and Health and Safety (Consultation with Employees) Regulations 1996, together with accompanying ACOP and HSE Guidance.

☞ **1(b) Statutes**
   Health and Safety at Work etc. Act 1974
☞ **1(c) Principal regulations**
   Health and Safety (Consultation with Employees) Regulations 1996
   Safety Representatives and Safety Committees Regulations 1977
☞ **1(d) Approved codes of practice**
   Safety representatives and safety committees
☞ **2(a) Health and safety in practice**
   Joint consultation
   Statements of health and safety policy
☞ **4. Health and Safety Glossary**
   Safety representative

L_{EQ} (Equivalent continuous sound level)

Where sound pressure levels fluctuate, an equivalent sound pressure level, averaged over a normal eight hour day.

Local effect

An effect on the body of exposure to a toxic substance which is at the initial point of contact, e.g. the skin, nose, throat, bladder, eyes.

1(e) HSE guidance notes
Workplace exposure limits
 2(b) Hazard checklists
Hazardous substances
 4. Health and Safety Glossary
Absorption
Acute effect
Chemical hazards
Chronic effect
Dose
Dose–response relationship
Hazardous substances
Health risk assessment
Health surveillance
Route of entry
Substances hazardous to health
Workplace exposure limit

Local exhaust ventilation (LEV) system

Mechanical exhaust ventilation systems designed to intercept airborne contaminants at the point of, or close to, the source of generation, directing the contaminant into a system of ducting connected to an extraction fan and filtration unit.

LEV systems incorporate:
* a hood, enclosure or inlet to collect the agent
* ductwork
* a filter or air-cleaning device
* a fan or other air-moving device
* further ductwork to discharge clean air to the external air.

LEV systems may be of the receptor, captor or low-volume high-velocity type (see individual entries).

 1(c) Principal regulations
Control of Lead at Work Regulations 2002
Control of Substances Hazardous to Health Regulations 2002

Ionising Radiations Regulations 1999
Workplace (Health, Safety and Welfare) Regulations 1992

☞ **1(d) Approved codes of practice**
    Control of substances hazardous to health

☞ **3(a) Tables and figures**
    Airborne contaminants: comparison of particle size ranges
    Local exhaust ventilation systems

☞ **4. Health and Safety Glossary**
    Containment
    Health risk assessment
    Risk control

Long-term exposure limit (LTEL)

LTELs for a wide range of chemical substances are listed in HSE Guidance Note EH40 *Workplace exposure limits*. They are concerned with the total intake of substances hazardous to health over long periods (8 hours), and are therefore appropriate for protecting against the effects of long-term exposure.

☞ **1(e) HSE guidance notes**
    Workplace exposure limits

Loss control

Any intentional management action directed at the prevention, reduction or elimination of the pure (non-speculative) risks of business.

A management system designed to reduce or eliminate all aspects of accidental loss that lead to waste of an organisation's assets.

☞ **2(a) Health and safety in practice**
    Accident costs
    Total loss control

Low voltage

This is a protective measure against electric shock The most commonly reduced low voltage system is the 110 volt centre point earthed system. With this system the secondary winding of the transformer providing the 110 volt supply is centre tapped to earth, thus ensuring that at no part of the 110 volt circuit can the voltage to earth exceed 55 volts.

☞ **1(c) Principal regulations**
    Electricity at Work Regulations 1989
☞ **1(d) Memorandum of guidance**
    Electricity at Work Regulations 1989
☞ **2(b) Hazard checklists**
    Electrical equipment
    Offices and commercial premises
☞ **4. Health and Safety Glossary**
    Intrinsically safe equipment
    Intrinsic safety

Lux

The metric unit of luminous flux or illuminance, which equates to lumens per square metre.

☞ **1(c) Principal regulations**
    Workplace (Health, Safety and Welfare) Regulations
☞ **1(d) Approved codes of practice**
    Workplace health, safety and welfare
☞ **3(a) Tables and figures**
    Average illuminances and minimum measured illuminances
    Maximum ratios of illuminance for adjacent areas
☞ **4. Health and Safety Glossary**
    Illuminance

Machinery guards

Safeguarding of machinery is achieved through a combination of physical guards and safety devices.

There are five main forms of machinery guard:
- a fixed guard i.e. a guard which has no moving parts associated with, or dependent upon, the mechanism of any machinery, and which, when in position, prevents access to a danger point or area
- an adjustable guard i.e. a guard incorporating an adjustable element which, once adjusted, remains in that position during a particular operation
- a distance guard i.e. a guard which does not completely enclose a danger point or area but which places it out of normal reach
- an interlocking guard i.e. a guard which has a movable part so connected with the machinery controls that:
    ○ the parts of the machinery causing danger cannot be set in motion until the guard is closed
    ○ the power is switched off and the motion braked before the guard can be opened sufficiently to allow access to the dangerous parts, and
    ○ access to the danger point or area is denied whilst the danger exists
- an automatic guard i.e. a guard which is associated with, and dependent upon, the mechanism of the machinery and operates so as to remove physically from the danger area any part of a person exposed to the danger.

See cross references for *Machinery hazards*

Machinery hazards

A person may be injured at machinery through:
- coming into contact with, or being trapped between, the machinery and any material in or at the machinery or any fixed structure

- being struck by, or becoming entangled in motion in, the machinery
- being struck by parts of the machinery ejected from it
- being struck by material ejected from the machinery (BS EN 292).

The principal hazards associated with machinery are:
- traps – reciprocating and shearing traps, and in-running nips
- entanglement – with unguarded rotating parts
- ejection – of items from machines
- contact – with, for instance, hot surfaces

☞ **1(a) Legal background**
Absolute (strict) liability
☞ **1(b) Statutes**
Health and Safety at Work etc. Act 1974
☞ **1(c) Principal regulations**
Lifting Operations and Lifting Equipment Regulations 1998
Provision and Use of Work Equipment Regulations 1998
☞ **1(d) Approved codes of practice**
Safe use of work equipment
☞ **2(a) Health and safety in practice**
Planned preventive maintenance
Risk assessment
☞ **2(b) Hazard checklists**
Maintenance work
Work equipment
☞ **4. Health and Safety Glossary**
CE marking
Contact hazard
Fail-safe
Fracture mechanics
Inspection (work equipment)
Intrinsically safe equipment
Machinery guards
Machinery safety devices
Non-mechanical hazards
Non-operational parts (machinery)

Operational parts (machinery)
Statutory examination
Statutory inspection

Machinery safety devices

Safety devices take the form of:
- trip devices i.e. a means whereby any approach by a person beyond the safe limit of working machinery causes the device to actuate and stop the machinery or reverse its motion, thus preventing or minimising injury at the danger point
- two-hand control devices, i.e. a device which requires both hands to operate the machinery controls, thus affording a measure of protection from danger only to the machinery operator and not other persons
- overrun devices, i.e. a device which, used in conjunction with a guard, is designed to prevent access to machinery parts which are moving by their own inertia after the power supply has been interrupted so as to prevent danger
- mechanical restraint devices, i.e. a device which applies mechanical restraint to a dangerous part of machinery which has been set in motion owing to failure of the machinery controls or other parts of the machinery, so as to prevent danger.

See cross references for *Machinery hazards*

Major injury

A major injury is classified as:
- any fracture, other than to the fingers, thumbs or toes
- any amputation
- dislocation of the shoulder, hip, knee or spine
- loss of sight (whether temporary or permanent)

- a chemical or hot metal burn to the eye or any penetrating injury to the eye
- any injury resulting from electric shock or electrical burn (including any electrical burn caused by arcing or arcing products) leading to unconsciousness or requiring resuscitation or admittance to hospital for more than 24 hours
- any other injury:
  - leading to hypothermia, heat-induced illness or to unconsciousness
  - requiring resuscitation, or
  - requiring admittance to hospital for more than 24 hours
- loss of consciousness caused by asphyxia or by exposure to a harmful substance or biological agent
- either of the following conditions which result from the absorption of any substance by inhalation, ingestion or through the skin:
  - acute illness requiring medical treatment
  - loss of consciousness
- acute illness which requires medical treatment where there is reason to believe that this resulted from exposure to a biological agent or its toxins or infected material.

[Reporting of Injuries, Diseases and Dangerous Occurrences Regulations 1995; Schedule 1]

☞ **1(c) Principal regulations**
   Reporting of Injuries, Diseases and Dangerous Occurrences Regulations 1995

☞ **2(a) Health and safety in practice**
   Accident costs
   Accident investigation procedures
   Major incidents
   Total loss control

☞ **3(a) Tables and figures**
   Reporting of Injuries, Diseases and Dangerous Occurrences Regulations – Reporting requirements

☞ **3(b) Forms**
   *Reporting of Injuries, Diseases and Dangerous Occurrences
      Regulations 1995:*
   (a) Report of an injury or dangerous occurrence (Form 2508)
   (b) Report of a disease (Form 2508A)
☞ **4. Health and Safety Glossary**
   Gas incident
   Reportable event

Manual handling operations

Any transporting or supporting of a load (including the lifting, putting down, pushing, pulling, carrying or moving thereof) by hand or bodily force.

[Manual Handling Operations Regulations 1992]

☞ **1(c) Principal regulations**
   Manual Handling Operations Regulations 1992
☞ **1(e) HSE guidance**
   Manual handling
☞ **2(b) Hazard checklists**
   Manual handling operations
☞ **3(a) Tables and figures**
   Manual handling – lifting and lowering
   Manual handling operations regulations – flow chart
☞ **3(b) Forms**
   Manual handling of loads – example of an assessment checklist
☞ **4. Health and Safety Glossary**
   Manual handling operations

Means of escape

A means of escape in case of fire is a continuous route by way of a space, room, corridor, staircase, doorway or other means of passage, along or through which persons can travel from wherever

they are in a building to the safety of the open air at ground level
by their own unaided efforts.

☞ **1(c) Principal regulations**
   Building Regulations 2000
   Regulatory Reform (Fire Safety) Order 2005
☞ **2(a) Health and safety in practice**
   Risk assessment
☞ **2(b) Hazard checklists**
   Fire safety
☞ **4. Health and Safety Glossary**
   Compartmentation
   Emergency lighting
   Fire instructions

Near miss

An unplanned and unforeseeable event that could have resulted
in death, human injury, property damage or other form of loss.

☞ **3(a) Tables and figures**
   Accident indices
   Accident ratios

Neutralisation

A control strategy for hazardous substances whereby a neutral-
ising compound is added to a highly dangerous compound,
e.g. acid to alkali, thereby reducing the immediate danger.
Many hazardous wastes are neutralised prior to transportation.

☞ **1(c) Principal regulations**
   Control of Substances Hazardous to Health Regulations 2002
☞ **1(d) Approved codes of practice**
   Control of substances hazardous to health
☞ **2(b) Hazard checklists**
   Hazardous substances

Noise-induced hearing loss (occupational deafness)

Exposure to noise at work may affect hearing in three ways:

- temporary threshold shift: a short-term effect, i.e. a temporary reduction in the ability to hear, which may follow exposure to excessive noise, such as that from rifle fire or certain types of machinery, such as chain saws
- permanent threshold shift: a permanent effect where the limit of tolerance is exceeded in terms of the duration and level of exposure to noise and individual susceptibility to noise
- acoustic trauma: a condition which involves sudden damage to the ear from short-term intense exposure or even from one single exposure, e.g. gun fire, major explosions.

Noise-induced hearing loss is a prescribed occupational disease.

 1(c) Principal regulations
Control of Noise at Work Regulations 2005

 2(a) Health and safety in practice
Dose record
Health records
Health surveillance

 **2(b) Hazard checklists**
Noise

 3(a) Tables and figures
Decibels (addition of)
Noise control methods
Noise control programme – Typical structure
Octave bands (standard range)

3(b) Forms
Noise exposure record

4. Health and Safety Glossary
Action levels, exposure action levels and exposure limit values
Audiometry
Dose
Impulse noise

L_{EQ} (equivalent continuous sound level)
Octave band analysis
Reduced time exposure (limitation)

Non-mechanical hazards

Those hazards associated with machinery but not arising from machinery motion, e.g. risk of burns from hot surfaces, contact with hazardous substances used in machines, exposure to machinery noise and airborne contaminants emitted from machines.

 **1(c) Principal regulations**
    Provision and Use of Work Equipment Regulations 1998
 **1(d) Approved codes of practice**
    Safe use of work equipment
 **2(a) Health and safety in practice**
    Planned preventive maintenance
    Risk assessment
 **2(b) Hazard checklists**
    Work equipment
 **4. Health and Safety Glossary**
    Contact hazard

Non-operational parts (machinery)

Those functional parts of machinery which convey power or motion to the operational parts, e.g. transmission machinery.

 **1(c) Principal regulations**
    Provision and Use of Work Equipment Regulations 1998
 **1(d) Approved codes of practice**
    Safe use of work equipment
 **2(a) Health and safety in practice**
    Planned preventive maintenance
    Risk assessment

2(b) Hazard checklists
   Work equipment

Occupational health

This is variously defined as:
- a branch of preventive medicine concerned with health problems caused by or manifest at work
- a branch of preventive medicine concerned with the relationship of work to health and the effects of work upon the worker.

 1(b) Statutes
   Health and Safety at Work etc. Act 1974
   Social Security Act 1975

 1(c) Principal regulations
   Chemicals (Hazard Information and Packaging for Supply) Regulations 2002
   Control of Asbestos at Work Regulations 2002
   Control of Lead at Work Regulations 2002
   Control of Noise at Work Regulations 2005
   Control of Substances Hazardous to Health Regulations 2002
   Control of Vibration at Work Regulations 2005
   Health and Safety (Display Screen Equipment) Regulations 1992
   Ionising Radiations Regulations 1999
   Management of Health and Safety at Work Regulations 1999
   Manual Handling Operations Regulations 1992
   Reporting of Injuries, Diseases and Dangerous Occurrences Regulations 1995

 1(d) Approved codes of practice
   Control of asbestos at work
   The management of asbestos in non-domestic premises
   Work with asbestos insulation, asbestos coating and asbestos insulation board
   Work with asbestos that does not normally require a licence
   Work with ionising radiation

Control of lead at work
The control of Legionella bacteria in hot water systems
Safe use of pesticides for non-agricultural purposes
Control of substances hazardous to health in the production of pottery
Control of substances hazardous to health
Control of substances hazardous to health in fumigation operations

☞ **1(e) HSE guidance notes**
A comprehensive guide to managing asbestos in premises [HS(G)227]
A Guide to the Work in Compressed Air Regulations 1989 [L102]
An introduction to local exhaust ventilation [HS(G)37]
Workplace exposure limits [EH40]

☞ **2(a) Health and safety in practice**
BS8800: Guide to Occupational Health and Safety Management Systems
Dose record
Health records
Health surveillance
Information and instruction
Risk assessment

☞ **3(a) Tables and figures**
Airborne contaminants: comparison of particle size ranges
Hazardous substances that can be revealed by medical analysis
Noise control methods
Optimum working temperatures
Reporting of Injuries, Diseases and Dangerous Occurrences Regulations – Reporting requirements

☞ **3(b) Forms**
Control of Substances Hazardous to Health Regulations – Health Risk Assessment
Noise exposure record
Occupational health:
(a) Pre-employment health questionnaire
(b) Health questionnaire

(c) Food handler's clearance certificate
(d) Fitness certificate
Reporting of Injuries, Diseases and Dangerous Occurrences
    Regulations 1995 – Report of a case of disease (Form
    2508A)

 4. Health and Safety Glossary

Absorption
Action levels, exposure action levels and exposure limit
    values
Acute effect
Air sampling (air monitoring)
Audiometry
Biological hazards
Biological monitoring
Carcinogen
Comfort ventilation
Dose
Dose–effect relationship
Dose–response relationship
Health risk assessment
Health surveillance
Impulse noise
Ionising radiation
Local effect
Occupational hygiene
Prescribed disease
Primary monitoring
Reduced time exposure (limitation)
Reportable disease
Risk avoidance
Route of entry
Secondary monitoring
Substances hazardous to health
Target organs and target systems
Threshold dose
Toxicity
Toxicological assessment
Toxicology

Occupational hygiene

The identification, measurement and control of contaminants and other phenomena, such as noise and radiation, which would otherwise have unacceptable adverse effects on the health of people exposed to them.

The four principal areas of occupational hygiene practice are:
- identification/recognition of the specific contaminant
- measurement, using an appropriate measuring technique
- evaluation against an existing standard e.g. Workplace Exposure Limits
- prevention or control of exposure.

See cross reference for *Occupational health*

Octave band analysis

A sound measurement technique which enables the way sound is distributed throughout the frequency spectrum to be identified. The sound is divided into octave bands and measured at the geometric centre frequency of each band.

Octave band analysis is used for assessing the risk of occupational deafness, in the analysis of machinery noise and specification of remedial measures, and in the specification of certain types of hearing protection.

Operational parts (machinery)

Those parts which perform the primary output function of a machine, namely the manufacture of a product or component, e.g. the chuck and drill bit on a vertical drill.

☞ **1(c) Principal regulations**
   Provision and Use of Work Equipment Regulations 1998
☞ **1(d) Approved codes of practice**
   Safe use of work equipment

2(a) Health and safety in practice
   Planned preventive maintenance
   Risk assessment
 2(b) Hazard checklists
   Work equipment

Organisational characteristics

Organisational characteristics which influence safety-related behaviour include:

- the need to produce a positive climate in which health and safety is seen by both management and employees as being fundamental to the organisation's day-to-day operations, that is, they must create a positive safety culture;
- the need to ensure that policies and systems which are devised for the control of risk from the organisation's operations take proper account of human capabilities and fallibilities;
- commitment to the achievement of progressively higher standards which is shown at the top of the organisation and cascaded through successive levels of same;
- demonstration by senior management of their active involvement, thereby galvanising managers throughout the organisation into action;
- leadership, whereby an environment is created which encourages safe behaviour.

[Reducing error and influencing behaviour – HS(G)48]

 1(c) Principal regulations
   Management of Health and Safety at Work Regulations 1999
 1(d) Approved codes of practice
   Management of health and safety at work
 1(e) HSE guidance notes
   Reducing error and influencing behaviour
   Successful health and safety management

Permit to work

A form of safe system of work operated where there is a high degree of foreseeable risk.

A formal safety control system designed to prevent accidental injury to personnel, damage to plant, premises and particularly when work with a foreseeably high hazard content is undertaken and the precautions required are numerous and complex.

2(b) Hazard checklists
Electrical equipment
Maintenance work
Radiation hazards
 4. Health and Safety Glossary
Confined space
Hot work
Job safety analysis

Personal protective equipment (PPE)

All equipment (including clothing affording protection against the weather) which is intended to be worn or held by a person at work and which protects him against one or more risks to his health and safety, and any addition or accessory designed to meet this objective.

[Personal Protective Equipment at Work Regulations 1992]
[Control of Substances Hazardous to Health Regulations 2002]

Any device or appliance designed to be worn or held by an individual for protection against one or more health and safety hazards; and shall also include:

- a unit constituted by several devices or appliances which have been integrally combined by the manufacturer for the protection of an individual against one or more potentially simultaneous risks;
- a protective device or appliance combined, separably or inseparably, with non-protective equipment worn or held by an individual for the execution of a specific activity; and
- interchangeable components which are essential to its satisfactory functioning and used exclusively for such equipment.

[Personal Protective Equipment Regulations 2002]

 1(c) Principal regulations
Construction (Head Protection) Regulations 1989
Control of Asbestos at Work Regulations 2002

Control of Lead at Work Regulations 2002
Control of Noise at Work Regulations 2005
Control of Substances Hazardous to Health Regulations 2002
Control of Vibration at Work Regulations 2005
Ionising Radiations Regulations 1999
Personal Protective Equipment at Work Regulations 1992

☞ **1(e) HSE guidance notes**
Personal protective equipment at work

☞ **2(b) Hazard checklists**
Noise
Personal protective equipment

☞ **3(a) Tables and figures**
Personal Protective Equipment at Work Regulations 1992 – Specimen risk survey table for the use of personal protective equipment

Place of safety

Generally interpreted as a location in the open air where people can freely walk away from a building and not be affected by heat or smoke from a fire in that building.

☞ **1(c) Principal regulations**
Regulatory Reform (Fire Safety) Order 2005

Prescribed disease

A disease may be prescribed if:
(a) it ought to be treated, having regard to its causes and incidence and other relevant considerations, as a risk of occupation and not a risk common to all persons; and
(b) it is such that, in the absence of special circumstances, the attribution of particular cases to the nature of the employment can be established with reasonable certainty.
[Social Security Act 1975]

Current requirements relating to prescribed occupational diseases are covered by the Social Security (Industrial Injuries) (Prescribed Diseases) Regulations 1985 and various amendments to these Regulations.

 1(b) The principal statutes
Social Security Act 1975

Primary monitoring

An area of occupational health practice dealing with the clinical observation of sick people who may seek advice and/or treatment.

 2(a) Health and safety in practice
Health records
Health surveillance

 3(a) Tables and figures
Hazardous substances that can be revealed by medical analysis

 3(b) Forms
Occupational health:
(a) Pre-employment health questionnaire
(b) Health questionnaire
(c) Food handler's clearance certificate
(d) Fitness certificate

 4. Health and Safety Glossary
Absorption
Acute effect
Chronic effect
Dose
Dose–response relationship
Health risk assessment
Health surveillance
Occupational health
Prescribed disease
Reportable disease

Product liability

An area of health and safety law concerned with both the criminal and civil liabilities of all those in the manufacturing chain towards consumers of their products. Criminal liability is covered in the HSWA (Sec 6) and other legislation, such as the Consumer Protection Act 1987. Injury sustained as a result of using a defective product could result in a civil claim against a defendant based on negligence.

Principal duties rest with designers, manufacturers and importers of products, secondary duties with wholesalers, retailers and other persons directly or indirectly involved in the supply chain.

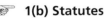 **1(b) Statutes**
   Health and Safety at Work etc. Act 1974
1(c) Principal regulations
   Chemicals (Hazard Information and Packaging for Supply) Regulations 2002
   Lifts Regulations 1997
   Pressure Systems Safety Regulations 2000
   Simple Pressure Vessels (Safety) Regulations 1991

Prohibition

A control strategy in accident and ill-health prevention exercised where there is no known form of operator protection available. This may entail prohibiting the use of a substance, system of work, operational practice or machine where the level of danger is very high.

An inspector is empowered to serve a prohibition notice where activities will, or may, involve a risk of serious personal injury.

1(a) Statutes
   Health and Safety at Work etc. Act 1974

Reduced time exposure (limitation)

A strategy directed to limiting the exposure of people to, for instance, noise or hazardous substances, by specifying the maximum exposure time permissible in any working period, e.g. eight hours. This strategy forms the basis for long-term and short-term exposure limits.

 1(c) Principal regulations
Control of Noise at Work Regulations 2005
Control of Vibration at Work Regulations 2005
Ionising Radiations Regulations 2005

Reduced voltage

A form of protection against electric shock, the most commonly used reduced voltage system being the 110 volt centre point earthed system. Here, the secondary winding of the transformer providing the 110 volt supply is centre tapped to earth, thereby ensuring that at no part of the 110 volt system can the voltage to earth exceed 55 volts.

 1(e) HSE guidance notes
Electricity at Work: Safe working practices
Electrical safety on construction sites

Reportable disease

Under the Reporting of Injuries, Diseases and Dangerous Occurrences Regulations (RIDDOR) 1995, certain diseases and conditions arising from work activities affecting a person at work, and listed in Schedule 3, must be reported by an employer to the relevant enforcing authority.

Reportable diseases are classified in the following groups:
   1. Conditions due to physical agents and the physical demands of work

2. Infections due to biological agents
3. Conditions due to substances.

☞ **1(c) Principal regulations**
Reporting of Injuries, Diseases and Dangerous Occurrences Regulations 1995
☞ **2(a) Health and safety in practice**
Accident investigation procedures
☞ **3(a) Tables and figures**
Reporting of Injuries, Diseases and Dangerous Occurrences Regulations – Reporting requirements
☞ **3(b) Forms**
Reporting of Injuries, Diseases and Dangerous Occurrences Regulations 1995
(b) Report of a case of disease (Form 2508A)

Reportable event

The Reporting of Injuries, Diseases and Dangerous Occurrences Regulations (RIDDOR) 1995 apply to 'events' which arise out of or in connection with work, namely all deaths, certain injuries resulting from accidents, instances of specified diseases and defined dangerous occurrences.

Where any of the events listed below arise out of work activities, it must be notified by quickest practicable means (e.g. telephone or fax), and subsequently reported (within 10 days) on the appropriate form, to the enforcing authority.

The events are:
- the death of any person at work as a result of an accident, whether or not they are at work
- someone who is at work suffering a major injury as a result of an accident arising out of or in connection with work
- someone who is not at work (e.g. a member of the public) suffers an injury as a result of an accident and is taken from the scene to a hospital for treatment, or if the accident happens at a hospital, suffers a major injury

- one of the list of specified dangerous occurrences takes place
- someone at work is unable to do their normal work for more than 3 days as a result of an injury caused by an accident at work
- the death of an employee, if this occurs some time after a reportable injury which led to the employee's death, but not more than 1 year afterwards
- a person at work suffers one of a number of specified diseases provided that a doctor diagnoses the disease and the person's job involves a specified work activity.

The duty to notify and report rests with the responsible person, i.e. the employer, a self-employed person or person in control of the premises.

A report must be made of the following circumstances:

- a conveyor of flammable gas through a fixed pipe distribution system or the filler, importer or supplier of liquefied petroleum gas (LPG) in a refillable container, must report if they learn that someone has died or suffered a major injury arising out of, or in connection with, that gas
- any registered installation business must report if it finds that there is, in any premises, a gas fitting or associated flue or ventilation arrangement which could be dangerous.

A responsible person must keep a record of the above events.

There is a defence available for a person to prove that he was not aware of the event requiring him to notify or send a report to the relevant authority, and that he had taken all reasonable steps to have such events brought to his notice.

 1(c) Principal regulations
> *Reporting of Injuries, Diseases and Dangerous Occurrences Regulations 1995*

 2(a) Health and safety in practice
> *Accident costs*
> *Accident investigation*
> *Major incidents*

☞ **3(a) Tables and figures**
  Reporting of Injuries, Diseases and Dangerous Occurrences Regulations – Reporting requirements

☞ **3(b) Forms**
  Reporting of Injuries, Diseases and Dangerous Occurrences Regulations 1995:
  (a) Report of an injury or dangerous occurrences (Form 2508)
  (b) Report of a case of disease (form 2508A).

Risk

Risk expresses the likelihood or probability that the harm from a particular hazard will be realised.

In relation to the exposure of an employee to a substance hazardous to health, means the likelihood that the potential for harm to health of a person will be attained under the conditions of use and exposure and also the extent of that harm.

[Control of Substances Hazardous to Health Regulations 2002]

☞ **1(c) Principal regulations**
  Control of Asbestos at Work Regulations 2002
  Control of Lead at Work Regulations 2002
  Control of Major Accident Hazards Regulations 1999
  Control of Noise at Work Regulations 2005
  Control of Substances Hazardous to Health Regulations 2002
  Control of Vibration at Work Regulations 2005
  Ionising Radiations Regulations 1999
  Lifting Operations and Lifting Equipment Regulations 1998
  Management of Health and Safety at Work Regulations 1999
  Manual Handling Operations Regulations 1992
  Personal Protective Equipment at Work Regulations 1992
  Regulatory Reform (Fire Safety) Order 2005

☞ **2(a) Health and safety in practice**
  Benchmarking

Risk avoidance

This strategy involves a conscious decision on the part of an employer to avoid completely a particular risk by, for instance, discontinuing or modifying the activities or operations that created the risk. An example might be the replacement of manual handling operations by a mechanical handling system.

 **1(a) Legal background**
Principles of Prevention
2(a) Health and safety in practice
Risk management

Risk control

This is one of the outcomes of the risk management process. Risk control may be through risk avoidance, risk retention, risk transfer or risk reduction (see individual entries).

 **1(a) Legal background**
Principles of prevention
2(a) Health and safety in practice
Risk management

Risk reduction

Risk reduction, as part of the risk management process, implies the implementation within an organisation of some form of loss control programme directed at protecting the organisation's assets (manpower, machinery, materials and money) from wastage caused by accidental loss.

Risk reduction strategies operate in two stages:
- collection of data on as many loss-producing incidents as possible and the installation of a programme of remedial action
- the collation of all areas where losses arise from loss-producing incidents, e.g. death, major injury, property damage, and the formulation of strategies directed at reducing these losses.

 1(a) Legal background
Principles of prevention
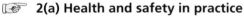 **2(a) Health and safety in practice**
Risk management

Risk retention

A risk management strategy whereby risk is retained within an organisation and any consequent loss is financed by the organisation. There are two features of risk retention:

Risk retention with knowledge
In this case a conscious decision is made to meet any resulting loss from within an organisation's resources. Decisions on which risks that should be retained can only be made after all the risks have been identified, measured and evaluated.

Risk retention without knowledge
This generally arises from a lack of knowledge of the existence of a risk or an omission to insure against that risk. Situations

where risks have not been identified and evaluated can result in this form of risk retention.

 2(a) Health and safety in practice
   Risk management

Risk transfer

Risk transfer implies the legal assignments of the costs of certain potential losses from one party to another, e.g. from an organisation to an insurance company.

 2(a) Health and safety in practice
   Risk management

Route of entry

There are three primary routes of entry of hazardous substances into the body, namely by inhalation, pervasion and ingestion. Secondary routes of entry include injection, inoculation and implantation.

 1(c) Principal regulations
   Chemicals (Hazard Information and Packaging for Supply) Regulations 2002
   Control of Asbestos at Work Regulations 2002
   Control of Lead at Work Regulations 2002
   Control of Substances Hazardous to Health Regulations 2002
 2(a) Health and safety in practice
   Health records
   Health surveillance
   Information and instruction
 2(b) Hazard checklists
   Hazardous substances

 **3(a) Tables and figures**

Airborne contaminants: comparison of particle size ranges

Categories of danger – Chemicals (Hazard Information and Packaging for Supply) Regulations 2002

Hazardous substances that can be revealed by medical analysis

Safety data sheets – obligatory headings

 **3(b) Forms**

Control of Substances Hazardous to Health Regulations 2002 – Health Risk Assessment

 **4. Health and Safety Glossary**

Absorption

Active monitoring

Acute effect

Aerosol

Air sampling (air monitoring)

Carcinogen

Chemical hazards

Chronic effect

Dilution ventilation

Dose

Dose–effect relationship

Dose–response relationship

Hazardous substances

Health risk assessment

Health surveillance

Local effect

Local exhaust ventilation

Long-term exposure limit

Primary monitoring

Secondary monitoring

Substances hazardous to health

Target organs and target systems

Threshold dose

Toxicity

Toxicological assessment

Toxicology

Workplace exposure limit

Safety culture

Both the HSE and CBI have provided guidance on the need for organisations to develop and promote the right safety culture.

The main principles involved, which involve the establishment of a safety culture, accepted and observed generally, are:

- the acceptance of responsibility at and from the top, exercised through a clear chain of command, seen to be actual and felt throughout the organisation
- a conviction that high standards are achievable through proper management
- setting and monitoring of relevant objectives/targets, based upon satisfactory internal information systems
- systematic identification and assessment of hazards and the devising and exercise of preventive systems which are subject to audit and review; in such approaches, particular attention is given to the investigation of error
- immediate rectification of deficiencies; and
- promotion and reward of enthusiasm and good results.

[Rimington, J.R. (1989) The Onshore Safety Regime, HSE Director General's Submission to the Piper Alpha Inquiry, December 1989]

A company wishing to improve its performance will need to judge its existing practices against a number of features essential to a sound safety culture, namely:

- leadership and commitment from the top which is genuine and visible; this is the most important feature
- acceptance that it is a long-term strategy which requires sustained effort and interest
- a policy statement of high expectations and conveying a sense of optimism about what is possible supported by adequate codes of practice and safety standards
- health and safety should be treated as other corporate aims, and adequately resourced
- it must be a line management responsibility
- 'ownership' of health and safety must permeate at all levels of the work force; this involves employee involvement, training and communication

- realistic and achievable targets should be set and performance measured against them
- incidents should be thoroughly investigated
- consistency of behaviour against agreed standards should be achieved by auditing and good safety behaviour should be a condition of employment
- deficiencies revealed by an investigation or audit should be remedied promptly
- management must receive adequate and up-to-date information to be able to assess performance.

[Developing a Safety Culture, CBI, 1989]

 2(a) Health and safety in practice
OHSAS 18001: A Pro-active Approach to Health and Safety Management
Successful Health and Safety Management

Safety propaganda

An important feature of communicating health and safety themes, hazards and messages to people, it may take the form of safety posters, films, demonstrations and exhibitions, directed at increasing awareness.

 1(c) Principal regulations
Health and Safety (Safety Signs and Signals) Regulations 1996
Safety Signs Regulations 1980
 2(a) Health and safety in practice
Information and instruction
Safety signs

Safety representative

The Safety Representatives and Safety Committees Regulations 1977 are concerned with the appointment by recognised trade unions of safety representatives, the functions of safety representatives and the establishment and operation of safety committees.

The functions of a safety representative include:

- to investigate potential hazards and dangerous occurrences and examine the causes of accidents
- to investigate health and safety complaints by the employees they represent
- to make representations to the employer on matters arising from investigations
- to make representations to the employer on general matters affecting health, safety and welfare
- to carry out inspections
- to represent employees in consultation with enforcement officers
- to receive information from enforcement officers
- to attend meetings of safety committees in their capacity as a safety representative in connection with the above functions.

☞ **1(c) Principal regulations**
Safety Representatives and Safety Committees Regulations 1977

☞ **1(d) Approved codes of practice**
Safety representatives and safety committees

☞ **4. Health and Safety Glossary**
Joint consultation

Secondary monitoring

An area of occupational health practice directed at controlling health hazards which have already been recognised, e.g. monitoring for occupational deafness by audiometry.

☞ **2(a) Health and safety in practice**
Health records
Health surveillance

4. Health and Safety Glossary
Absorption
Action levels, exposure levels and exposure limit values
Audiometry
Biological monitoring
Health surveillance
Occupational health
Primary monitoring

Segregation

This is a strategy aimed at controlling the hazards arising from toxic substances and certain physical hazards, such as noise and radiation. Segregation may take a number of forms:
- segregation by distance (separation)
- segregation by age
- segregation by time
- segregation by sex
- segregation by physiological criteria.

 2(a) Health and safety in practice
Health surveillance
 4. Health and Safety Glossary
Health risk assessment
Health surveillance
Occupational health
Reduced time exposure (limitation)

Statutory examination

With reference to Regulation 32 of the Provision and Use of Work Equipment Regulations 1998 thorough inspection in relation to a thorough examination
- means a thorough examination by a competent person
- includes testing the nature and extent of which are appropriate for the purpose described in the regulation.

1(c) Principal regulations
Provision and Use of Work Equipment Regulations 1998
 2(a) Health and safety in practice
Competent persons

Statutory inspection

With reference to Regulation 6 of the Provision and Use of
Work Equipment Regulations 1998
- such visual or more rigorous inspection by a competent
  person as is appropriate for the purpose described in the
  regulation
- where it is appropriate to carry out testing for the purpose, includes testing the nature and extent of which are
  appropriate for the purpose.

 1(c) Principal regulations
Provision and Use of Work Equipment Regulations 1998
 2(a) Health and safety in practice
Competent persons

Substance hazardous to health

Any substance (including any preparation):
- (a) which is listed in Part 1 of the Approved Supply List as dangerous for supply within the meaning of the Chemicals
  (Hazard Information and Packaging for Supply) Regulations and for which an indication of danger specified for
  the substance in Part V of that list is very toxic, toxic, harmful, corrosive or irritant;
- (b) for which the HSC has approved a maximum exposure
  limit or an occupational exposure standard;
- (c) which is a biological agent;
- (d) dust of any kind, except dust which is a substance within
  paragraph (a) or (b) above, when present at concentration

in air equal to or greater than:
- 10 mg/m^3, as a time-weighted average over an 8-hour period of total inhalable dust; or
- 4 mg/m^3, as a time-weighted average over an 8-hour period of respirable dust.

(e) which, not being a substance falling within sub-paragraphs (a) to (d), because of its chemical or toxicological properties and the way it is used or is present at the workplace creates a risk to health.

[Control of Substances Hazardous to Health Regulations 2002]

 1(c) Principal regulations
Control of substances Hazardous to Health Regulations 2002

 1(d) Approved codes of practice
Control of substances hazardous to health
Control of substances hazardous to health in fumigation operations

 2(b) Hazard checklists
Hazardous substances

 3(c) Tables and figures
Categories of danger – Chemicals (Hazard Information and Packaging for Supply) Regulations 2002

Substitution

A prevention strategy whereby a less hazardous substance, process or work activity is substituted for a more dangerous one.

 Health and Safety Glossary
Chemical hazards
Hazardous substances

Target organ and target system

Certain toxic substances have a direct or indirect effect on specific body organs (target organs) and body systems (target systems).

Target organs include the liver, lungs, bladder, brain and skin. Target systems include the respiratory system, circulatory system, lymphatic system and reproductive system.

 1(b) Hazard checklists
   Hazardous substances
 4. Health and Safety Glossary
   Absorption
   Dose
   Dose–effect relationship
   Dose–response relationship
   Occupational health
   Toxicity
   Toxicological assessment
   Toxicology

Threshold dose

A concentration of an offending agent in the body above which an adverse body response will take place.

 4. Health and Safety Glossary
   Absorption
   Dose
   Dose–effect relationship
   Dose–response relationship
   Toxicity
   Toxicological assessment
   Toxicology
   Workplace exposure limits

Toxicity

The ability of a chemical molecule to produce injury once it reaches a susceptible site in or on the body.

3(a) Tables and figures
> *Categories of danger – Chemicals (Hazard Information and Packaging for Supply) Regulations 2002*

 4. Health and Safety Glossary
> *Absorption*
> *Dose*
> *Dose–effect relationship*
> *Threshold dose*
> *Toxicological assessment*
> *Workplace exposure limits*

Toxicological assessment

The collection, assembly and evaluation of data on a potentially toxic substance and the conditions of its use, in order to determine:

- the danger to human health
- systems for preventing or controlling the danger
- the detection and treatment of overexposure and,
- where such information is insufficient, the need for further investigation.

The following factors should be considered in toxicological assessment:

- the name of the substance, including any synonyms
- a physical and/or chemical description of the substance
- information on potential exposure situations
- details of occupational exposure limits
- general toxicological aspects, such as
  - the route of entry into the body
  - the mode of action in or on the body
  - signs and symptoms
  - diagnostic tests
  - treatment, and
  - disability potential.

 1(c) Principal regulations
> *Chemicals (Hazard Information and Packaging for Supply) Regulations 2002*

Toxicology

The study of the body's responses to toxic substances.

See cross references for *Toxicological assessment*

Welfare amenity provisions

Detailed requirements relating to the provision and main-tenance of welfare amenities are covered in the Workplace (Health, Safety and Welfare) Regulations 1992 and accom-panying Approved Code of Practice.

Duties on employers relating to welfare amenities include the provision of:
- suitable and sufficient sanitary conveniences and wash-ing facilities
- an adequate supply of wholesome drinking water
- accommodation for clothing
- facilities for changing clothing
- facilities for rest and the taking of meals.

Schedule 1 to the regulations covers provisions with regard to the number of sanitary fitments and washing facilities to be provided.

 1(b) Statutes
Health and Safety at Work etc. Act 1974
 1(c) Principal regulations
Construction (Health, Safety and Welfare) Regulations 1996
Control of Asbestos at Work Regulations 2002
Control of Lead at Work Regulations 2002
Control of Substances Hazardous to Health Regulations 2002
Ionising Radiations Regulations 1999
Workplace (Health, Safety and Welfare) Regulations 1992
 3(a) Tables and figures
Water closets and urinals for men
Water closets and wash station provision

Workplace exposure limits

HSE Guidance Note EH 40 'Workplace exposure limits', which is revised on a regular basis, lists details of occupational exposure limits (OELs) set under the COSHH Regulations in order to help protect the health of workers. The list of WELs, unless otherwise stated, relates to personal exposure to substances hazardous to health in the workplace only. They are intended to be used for normal working conditions in factories or other workplaces.

Workplace exposure limits (WELs) are concentrations of hazardous substances in the air, averaged over a specified period of time referred to as a time-weighted average (TWA). Two time periods are used:

- long term (8 hours)
- short term (15 minutes).

Long-term and short-term exposure limits

Effects of exposure to substances hazardous to health vary considerably depending upon the nature of the substance and the pattern of exposure. Some effects require prolonged or accumulated exposure.

Long-term (8-hour TWA) exposure limits (LTELs) are intended to control such effects by restricting the total intake by inhalation over one or more work shifts, depending upon the length of the shift. Other effects may be seen after brief exposure.

Short-term exposure limits (usually 15 minutes) (STELs) may be applied to control these effects. They are set to help prevent effects, such as eye irritation, which may occur following exposure for a few minutes. For those substances for which no STEL is specified, it is recommended that a figure of three times the LTEL be used as a guideline for controlling short-term peaks in exposure.

Some workplace activities give rise to frequent short (less than 15 minutes) periods of high exposure which, if averaged over time, do not exceed either an 8-hour TWA or a 15-minute TWA. Such exposures have the potential to cause harm and should be subject to reasonably practicable means of control unless a 'suitable and sufficient' risk assessment shows no risk to health from such exposures.

Regulation 7(7) of the COSHH Regulations states that:

'Without prejudice to the requirement to prevent exposure, control of exposure to a substance hazardous to health shall only be treated as adequate if:

(a) the principles of good practice for the control of exposure to substances hazardous to health set out in Schedule 2A are applied;

(b) any *workplace exposure limit* is not exceeded; and

(c) for a substance:

 (i) which carries the risk phrase R45, R46 or R49, or for a substance or process which is listed in Schedule 1; or

(ii) which carries the risk phrase R42 or R42/43, or which is listed in section C of HSE publication *Asthmagen? Critical assessments of the evidence for agents implicated in occupational asthma* as updated from time to time, or any other substance which the risk assessment has shown to be a potential source of occupational asthma, exposure is reduced to as low a level as is reasonably practicable.'

WELs are listed in Schedule 1 to the Guidance Note.

However, the absence of a substance from the list of WELs does not indicate that it is safe. For these substances, exposure should be controlled to a level to which nearly all the working population could be exposed, day after day at work, without adverse effects on health.

Absorption through the skin

For most substances, the main route of entry into the body is by inhalation and the exposure limits given relate solely to exposure by this route. However, some substances have the ability to penetrate intact skin and become absorbed into the body, thus contributing to systemic toxicity. These substances are marked in the Table of values with an 'Sk' notation. The Advisory Committee on Toxic Substances (ACTS) has agreed the following criteria for assigning this notation.

The 'Sk' notation is assigned in cases where the available data or experience (or predictions made in the absence of actual data) suggest that exposure via the dermal route may:

(a) make a substantial contribution to the body burden (when compared to the contribution attributable to inhalation exposure at the WEL); and

(b) cause systemic effects, so that conclusions about exposure and health effects based solely on airborne concentration limits may be incompatible.

Units of measurement

In WELs, concentrations of airborne particles (fume, dust, etc.) are usually expressed in $mg\,m^{-3}$. In the case of dusts, the limits in the table refer to the 'inhalable' fraction unless specifically indicated as referring to the 'respirable' fraction. WELs for volatile substances are usually expressed in both parts per million by volume (p.p.m.) and milligrams per cubic metre ($mg\,m^{-3}$). For these substances, limits are set in ppm, and a conversion to $mg\,m^{-3}$ is calculated.

European occupational exposure limits

The exposure limits listed in Table 1 are all British limits set under the COSHH Regulations. In some cases, these also reflect a European limit applicable in all EU Member States. These limits are currently known as *Indicative Occupational Exposure Limit Values (IOELVs)*.

IOELVs are health-based limits set under the Chemical Agents Directive (98/24/EC). The European Commission is advised on limits by its Scientific Committee on Occupational Exposure Limits (SCOEL). This committee evaluates the scientific information on hazardous substances and makes recommendations for the establishment of an IOELV. IOELVs are listed in Directives which Member States are obliged to implement by introducing national limits for the substances listed.

PART 5
Appendices

Appendix A: Accredited training courses in occupational health and safety

The following professional institutions promote and accredit training courses.

Institution of Occupational Safety and Health (IOSH)

Directing Safely
Managing Safely
Managing Client/Contractor Relationships
Managing Risk
Managing Safely in Policing Services
Managing with Environmental Responsibilities
Working Safely
Working with Environmental Responsibilities
Health Care: Working Safely
Health Care: Risk and Safety Management

Chartered Institute of Environmental Health (CIEH)

Foundation Certificate in Health and Safety
Risk Assessment: Principles and practice
Supervising Health and Safety
Advanced Certificate in Health and Safety in the Workplace
Principles of COSHH
Principles of Manual Handling

National Examination Board in Occupational Safety and Health (NEBOSH)

Diploma Parts 1 and 2 in Occupational Safety and Health
Level 4 Diploma in Occupational Safety and Health
Specialist Diploma in Environmental Management
National General Certificate in Occupational Safety and Health
International General Certificate in Occupational Safety and Health
National Certificate in Construction Safety and Health

Royal Institute of Public Health (RIPH)

Foundation Certificate in Health and Safety in the Workplace

Royal Society of Health (RSH)

Foundation Certificate in Health and Safety in the Workplace
Advanced Diploma in Health and Safety at Work

Appendix B: Documentation and record keeping requirements

Current health and safety legislation places considerable emphasis on the documentation of policies, procedures and systems of work and the maintenance of certain records.

The following are some of the documents and records that are required to be produced and maintained or may be required to be shown as evidence of compliance with health and safety legislation.

- Statement of Health and Safety Policy (Health and Safety at Work etc. Act 1974)
- Risk assessments in respect of:
  - workplaces [Management of Health and Safety at Work Regulations 1999 and Workplace (Health, Safety and Welfare) Regulations 1992]
  - work activities [Management of Health and Safety at Work Regulations 1999 and Workplace (Health, Safety and Welfare) Regulations 1992]
  - work groups [Management of Health and Safety at Work Regulations]
  - new or expectant mothers [Management of Health and Safety at Work Regulations 1999]
  - young persons [Management of Health and Safety at Work Regulations 1999]
  - work equipment [Provision and Use of Work Equipment Regulations 1998]
  - personal protective equipment [Personal Protective Equipment Regulations 1992]
  - manual handling operations [Manual Handling Operations Regulations 1992]
  - display screen equipment [Health and Safety (Display Screen Equipment) Regulations 1992]

- ○ substances hazardous to health [Control of Substances Hazardous to Health Regulations 2002]
- ○ significant exposure to lead [Control of Lead at Work Regulations 2002]
- ○ noise at or above a lower exposure action value, i.e
  - (a) a daily or weekly personal noise exposure of 80 dB (A-weighted); and
  - (b) a peak sound pressure of 135 dB (C-weighted) [Control of Noise at Work Regulations 2005]
- ○ work liable to expose employees to vibration [Control of Vibration at Work Regulations 2005]
- ○ before a radiation employer commences a new activity involving work with ionising radiation [Ionising Radiations Regulations 1999]
- ○ the presence or otherwise of asbestos in non-domestic premises [Control of Asbestos at Work Regulations 2002]
- ○ work at height [Work at Height Regulations 2005]
- ○ where a dangerous substance is or is liable to be present at the workplace [Dangerous Substances and Explosive Atmospheres Regulations 2002]
- Safe systems of work, including permits to work and method statements [Health and Safety at Work etc Act 1974, Construction (Design and Management) Regulations 1994, Confined Spaces Regulations 1997, Control of Substances Hazardous to Health Regulations 2002, Work in Compressed Air Regulations 1996]
- Pre-tender stage health and safety plan and construction phase health and safety plan [Construction (Design and Management) Regulations 1994]
- Planned preventive maintenance schedules [Workplace (Health, Safety and Welfare) Regulations 1992 and Provision and Use of Work Equipment Regulations 1998]
  - ○ Cleaning schedules [Workplace (Health, Safety and Welfare) Regulations 1992]
- Written scheme of examination for specific parts of an installed pressure system or of a mobile system and the

last report relating to a system by a competent person [Pressure Systems Safety Regulations 2000]

- Written plan of work identifying those parts of a premises where asbestos is or is liable to be present in a premises and detailing how that work is to be carried out safely and without risk to health [Control of Asbestos at Work Regulations 2002]
- Records of examinations and tests of exhaust ventilation equipment and respiratory protective equipment and of repairs carried out as a result of those examinations and tests [Control of Lead at Work Regulations 1999, Control of Substances Hazardous to Health Regulations 2002 and Control of Asbestos at Work Regulations 2002]
- Record of air monitoring carried out in respect of:
  - specified substances or processes; and
  - lead
  - asbestos

  [Control of Substances Hazardous to Health Regulations 2002, Control of Lead at Work Regulations 1998 and Control of Asbestos at Work Regulations 2002]
- Record of examination of respiratory protective equipment [Ionising Radiations Regulations 1999]
- Records of air monitoring in cases where exposure to asbestos is such that a health record is required to be kept [Control of Asbestos at Work Regulations 2002]
- Personal health records [Control of Lead at Work Regulations 2002, Ionising Radiations Regulations 1999, Control of Substances Hazardous to Health Regulations 2002 and Control of Asbestos at Work Regulations 2002]
- Personal dose records [Ionising Radiations Regulations 1999]
- Record of quantity and location of radioactive substances [Ionising Radiations Regulations 1999]
- Record of investigation of certain notifiable occurrences involving release or spillage of a radioactive substance [Ionising Radiations Regulations 1999]

- Record of suspected overexposure to ionising radiation during medical exposure [Ionising Radiations Regulations 1999]
- Major Accident Prevention Policy [Control of Major Accident Hazards Regulations 1999]
- Off-Site Emergency Plan [Control of Major Accident Hazards Regulations 1999]
- Declaration of conformity by the installer of a lift and the manufacturer of a safety component for a lift together with any technical documentation or other information in relation to a lift or safety component required to be retained under the conformity assessment procedure [Lifts Regulations 1997]
- Declaration of conformity by the manufacturer of pressure equipment and assemblies (as defined) together with technical documentation or other information in relation to an item of pressure equipment and assemblies required to be retained under the conformity assessment procedure used [Pressure Equipment Regulations 1999]
- Any technical documentation or other information required to be retained under a conformity assessment procedure and a periodic inspection procedure [Transportable Pressure Vessels Regulations 2001]
- Procedures for serious and imminent danger and for danger areas [Management of Health and Safety at Work Regulations 1999]
- Emergency procedure to protect the safety of employees from an accident, incident or emergency related to the presence of a dangerous substance at the workplace [Dangerous Substances and Explosive Atmospheres Regulations 2001]
- Contingency plan in the event of a radiation accident [Ionising Radiations Regulations 1999]
- Local rules in respect of controlled areas and supervised areas [Ionising Radiations Regulations 1999]
- Written arrangements for non-classified persons [Ionising Radiations Regulations 1999]

Appendix C: Useful publications and information sources

Examination syllabuses and reports

National Examination Board in Occupational Safety and Health
Institution of Occupational Safety and Health
Chartered Institute of Environmental Health
Royal Society of Health
Royal Institute of Public Health
British Safety Council
Royal Society for the Prevention of Accidents

Health and safety books

General health and safety information

Dewis M. & Braune J. (2005): *Tolley's Health and Safety at Work Handbook*: LexisNexis/Tolley Publishing

Hughes P. & Ferret E. (2005): *Introduction to Health and Safety at Work*, 2nd edition: Elsevier/Butterworth-Heinemann

Fuller C. & Vassie L. (2004): *Health and Safety Management*: Pearson Education

Health and Safety Executive (1999): *Essentials of Health and Safety at Work*: HSE Books

Holt, Allan St.J. (2002): *Principles of Health and Safety at Work*, 6th edition: IOSH Publications Ltd

O'Donnell M. P. (2001): *Health Promotion in the Workplace*: Thomson Delmar Learning

Ridley J. (2004): *Health and Safety in Brief*, 3rd edition: Elsevier/Butterworth Heinemann

Ridley J. & Channing J. (2003): *Safety at Work*, 6th edition: Elsevier/Butterworth Heinemann

Stranks J. (2003): *A Manager's Guide to Health and Safety at Work*: Kogan Page

Stranks J. (2003): *Health and Safety for Management*: Highfield Publishing

Stranks J. (2005): *The Handbook of Health and Safety Practice*, 7th edition: Pearson Education

Case studies

Kletz T. A. (2001): *Learning from Accidents*, 3rd edition: Elsevier/Butterworth-Heinemann

Health and safety law

Barret B. & Howells R. (2000): *Occupational Health and Safety Law: Cases and Materials*: Cavendish Publishing

Goodman M. J. et al. (1999): *Encyclopedia of Health and Safety at Work*: Gee Publishing

Kloss D. (2005): *Occupational health law,* 4th edition*:* Blackwell Publishing

Reeve P. et al. (2000): *Health and Safety Competent Person's Handbook*: Gee Publishing

Stranks J. (2005): *Health and Safety Law*, 5th edition: Pearson Education

Occupational health and hygiene

Ashton I. & Gill F. (1999): *Monitoring for Health Hazards at Work*, 3rd edition: Blackwell Publishing

Baxter P. et al. (2000): *Hunter's Diseases of Occupations*, 9th edition: Hodder Arnold

Cooper C. & Clarke S. (2003): *Managing the Risk of Workplace Stress: Health and Safety Hazards*: Taylor & Francis/Routledge

Gardiner K. & Harrington J. M. (2005): *Occupational Hygiene*, 3rd edition: Blackwell Publishing

Harrington J. M. et al. (1998): *Occupational Health (Pocket Consultant series)*, 4th edition: Blackwell Publishing

Harris R. et al. (2001): *Patty's Industrial Hygiene and Toxicology*, 5th edition: John Wiley & Sons

Hartley C. (2000): *Health and Safety: Hazardous Agents*: IOSH Publications Ltd

Kloss D. (2005): *Occupational Health Law*, 4th edition: Blackwell Publishing

Sadhra S. & Rampal K. (1999): *Occupational Health: Risk Assessment and Management*: Blackwell Publishing

Stranks J. (2005): *Stress at Work: Management and Prevention*: Elsevier/Butterworth-Heinemann

Waldron H. A. & Edling C. (1997): *Occupational Health Practice*: Hodder Arnold

Williams N. & Harrison R. (2004): *Atlas of Occupational Health and Diseases*: Hodder Arnold

Risk assessment and management

Bateman M. (2003): *Tolley's Practical Risk Assessment Handbook*, 4th edition: LexisNexis/Elsevier

Boyle T. (2003): *Health and Safety: Risk Management*, 2nd edition: IOSH Publications Ltd

Sadhra S. & Rampal K. (1999): *Occupational Health: Risk Assessment and Management*: Blackwell Publishing

Construction safety

Holt, Allan St.J (2001): *Principles of Construction Safety*: Blackwell Publishing

Hughes P. & Ferrett E. (2004): *Introduction to Health and Safety in Construction*: Elsevier/Butterworth-Heinemann

Fire safety

HMSO (1999): *Fire safety: an employer's guide*: HMSO

Thomson N. G. (2001): *Fire Hazards in Industry*: Elsevier/Butterworth-Heinemann

Specific topics

Kroemer K. H. E. & Grandjean E. (1997) *Fitting The Task To The Human: A Textbook Of Occupational Ergonomics*, 5th Edition: Taylor & Francis

Lakha R. & Moore T. (2004): *Tolley's Handbook of Disaster and Emergency Management: Principles and Practice*: LexisNexis/ Elsevier

Ridley J. & Pearce D. (2005): *Safety with Machinery*, 2nd edition: Elsevier/Butterworth-Heinemann

Health and safety periodicals

Safety & Health Practitioner

Official magazine of the Institution of Occupational Safety and Health (IOSH), published by CMP. Also online at: www.shponline.co.uk

CMP Information, 245 Blackfriars Road, London SE1 9UY, UK

Subscriptions:

Tel: +44 (0)1635 588890   Fax: +44 (0)1635 868594
    email: shpcirculation@cmpinformation.com

Editorial:

Tel: +44 (0)20 7921 8046   Fax: +44 (0)20 7921 8058
    email: shpeditor@cmpinformation.com

Health and Safety at Work magazine

Published by LexisNexis Butterworths.

Journals and Magazines, 2 Addiscombe Road, Croydon, Surrey CR9 5AF, UK

Subscriptions:

Tel: +44 (0)20 86869141   Fax: +44 (0)20 86861910

Editorial:

Tel: +44 (0)20 82121938   email: hsw@lexisnexis.co.uk

Occupational Safety & Health

Official journal of the Royal Society for the Prevention of Accidents (RoSPA)

Press and Periodicals, Edgbaston Park, 353 Bristol Road, Birmingham B5 7ST, UK

Sales:

Tel: +44 (0)870 777 2227   email: sales@rospa.com

Editorial:
Tel: +44 (0)121 248 2000   Fax: +44 (0)121 248 2001

Hazards
Official journal of the Trades Union Congress (TUC). Also
    online at www.hazards.org
PO Box 199, Sheffield S1 4YL, UK
Subscription:
Tel: +44 (0)114 235 2074   email: sub@hazards.org
Editorial:
Tel: +44 (0)114 201 4265   email: editor@hazards.org

Environmental Health Journal
Official Journal of the Chartered Institute of Environmental
    Health
Chadwick Court, 15 Hatfields, London SE1 8DJ, UK
Subscription:
Tel: +44 (0)20 7827 5882   email: j.godden@chgl.com
Editorial:
Tel: +44 (0)20 7928 6006   email: ehj@chgl.com

Safety Management
Official journal of the British Safety Council
70 Chancellors Road, London W6 9RS, UK
Subscription:
Tel: +44 (0)20 8741 1231   Fax: +44 (0)20 8741 4555
    email: subscriptions@britsafe.org
Editorial:
Tel: +44 (0)20 8741 1231   Fax: +44 (0)20 8741 4555
    email: publications@britsafe.org

HSC and HSE publications and information services

HSE Books
HSE Books is the publishing arm of the Health and Safety
Executive and the mail order and warehousing service for the

distribution of HSC/HSE publications. It distributes both priced and free publications and also operates a number of subscription services.

HSE Books issue a catalogue at regular intervals.

Contact HSE Books, PO Box 1999, Sudbury, Suffolk CO10 2WA Tel +44 (0)1787 881165

The catalogue can also be found online at the HSE Bookfinder website at www.hsebooks.com. This website provides the on-line system for selecting and ordering publications from HSE Books. The website provides a wide range of information including:

- free leaflets
- general occupational safety and health
- health and safety topics
- HSE research
- People in the workplace.

The website is updated on a weekly basis to take into account the addition of new publications and titles.

HSE website

The HSE has a website on the Internet. This contains information about the objectives of the HSE, how to contact the HSE, how to complain, recent press releases and research and current initiatives. Information about risks at work and information about different workplaces is also available.

A feedback facility is available to enable organisations and individuals to post enquiries and suggestions to the HSE electronically.

The URL for accessing the HSE home page is www.hse.gov.uk.

hsedirect

Developed by the HSE in partnership with LexisNexis, hsedirect is an on-line information service providing instant access to the latest legislation, ACOPs and HSE guidance, EU Directives, British Standard summaries, case summaries and HSE forms. The site also contains daily news, HSE press releases, a health and safety events diary and useful contact details.

The home page can be contacted through www.hsedirect.com. with further enquiries through enquiries@hsedirect.com (tel. 0845 300 3142).

HSELINE

HSELINE is a computer database of bibliographic references to published documents on health and safety at work. It contains over 230 000 references and over 9000 additions are made each year.

For further information on how to access HSELINE contact:

Dialtech, 148 Darland Avenue, Gillingham, Kent ME7 3AS
Tel: 01634 574592
email: einsuk@aol.com

or

Dialog, Thompson Corporation, Palace House, 3 Cathedral Street, London SE1 9DE Tel: 0207 940 6900 Fax: 0207 940 6800

HSE Infoline

Infoline is the HSE's public enquiry contact centre. It's a 'one-stop shop', providing rapid access to the HSE's wealth of health and safety information, and access to expert advice and guidance.

HSE Infoline can be contacted by telephone (08701 545500), Minicom (02920 808537) Fax (02920 859260), email (hseinformationservices@natbrit.com) or by post (HSE Infoline, Caerphilly Business Park, Caerphilly CF83 3GG).

HSE electronic journals
HSE electronic journals include:
- *Biological Agents Bulletin*
- *Local Authority Unit Newsletter*
- *Radiation Protection News*
- *Safety Statistics Bulletin*
- *Site Safe News*
- *Toxic Substances Bulletin*

HSC newsletter
This bi-monthly publication provides a single source of information for all those affected by health and safety issues, whether managers, shop floor workers, safety officers or safety representatives (available through HSE Books).

HSE news bulletin
A weekly compilation of all press releases issued by the HSE's press office on a variety of subjects relating to health and safety in the workplace (available through HSE Books).

HSE information series
This series provides guidance on a very broad range of topics, including:

A guide to the Construction (Health, Safety and Welfare) Regulations 1996
An introduction to health and safety
A short guide to managing asbestos in premises
A short guide to the Personal Protective Equipment at Work Regulations 1992
Back pain: managing back pain in the workplace
Basic advice on first aid at work
Buying new machinery
Checkouts and musculoskeletal disorders
Computer control: a question of safety

Consulting employees on health and safety
Contained use of genetically modified organisms
COSHH: a brief guide to the Regulations
COSHH and section 6 of the Health and Safety at Work Act
Driving at work: managing work-related road safety
Electrical safety and you
Electric storage batteries: safe charging and use
Emergency action for burns
Employers' Liability (Compulsory Insurance) Act 1969: A guide
    for employers and their representatives
Ergonomics at work
Fire and explosion: how safe is your workplace?
First aid at work
First aid at work: your questions answered
Five steps to information, instruction and training
Five steps to risk assessment
Five steps to successful health and safety management
Gas appliances: get them checked: keep them safe
Getting to grips with manual handling
Grin and wear it
Hand-arm vibration
Health and safety benchmarking
Health and safety regulation: a short guide
Health risks from hand-arm vibration: advice for employers
Health surveillance in noisy industries: advice for employers
Homeworking
If the task fits: ergonomics at work
Keep the noise down
Legionnaires' disease
Lighten the load: guidance for employers on musculoskeletal
    disorders
Listen up!
Maintaining portable electrical equipment in offices and other
    low risk environments
Managing asbestos in workplace buildings
Managing crowds safely
Managing health and safety: five steps to success
Managing vehicle safety in the workplace

Manual handling: a short guide for employers

Manual handling assessment charts

Noise at work: advice for employers

Noise in construction: further guidance on the Noise at Work Regulations 1989

Officewise

Passive smoking at work

Preventing dermatitis at work

Preventing slips, trips and falls at work

Protecting your health at work

Read the label: how to find if chemicals are dangerous

Reduce risks – cut costs: the real costs of accidents and ill health at work

Respiratory sensitisers

RIDDOR explained: Reporting of Injuries, Diseases and Dangerous Occurrences Regulations 1995

Safety in electrical testing

Safe working with flammable substances

Signpost to the Health and Safety (Safety Signs and Signals) Regulations 1996

Silica dust and you

Starting your business: guidance on preparing a health and safety policy document for small firms

Tackling work-related stress: a guide for employees

The complete idiot's guide to CHIP

The Noise at Work Regulations: a brief guide to the requirements for controlling noise at work

The right start: work experience for young people

Training woodworking machinists

Understanding ergonomics at work

Upper limb disorders; assessing the risks

Using work equipment safely

Violence at work: a guide for employers

Welfare at work

What your doctor needs to know: your work and your health

Working alone in safety: controlling the risks of solitary work

Working with VDUs

Workplace health, safety and welfare

Appendix D: Professional organisations

Professional Organisations

Institution of Occupational Safety and Health (IOSH)
The Grange, Highfield Drive, Wigston, Leicester LE18 1NN, UK
Tel: +44 (0)116 257 3100   Fax: +44 (0)116 257 3101
email: enquiries@iosh.co.uk   Website: www.iosh.co.uk

British Occupational Hygiene Society (BOHS)
5/6 Melbourne Business Court, Millennium Way, Pride Park, Derby, DE24 8LZ, UK
Tel: +44 (0)1332 298101   Fax: +44 (0)1332 298099
email: admin@bohs.org   Website: www.bohs.org

Chartered Institute of Environmental Health (CIEH)
Chadwick Court, 15 Hatfields, London SE1 8DJ, UK
Tel: +44 (0)20 7928 6006   Fax: +44 (0)20 7827 5862
email: info@cieh.org   Website: www.cieh.org

Ergonomics Society
Elms Court, Elms Grove, Loughborough LE11 1RG, UK
Tel: +44 (0)1509 234904   Fax: +44 (0)1509 235666
email: ergsoc@ergonomics.   Website: www.
org.uk   ergonomics.org.uk

Institute of Occupational and Environmental Medicine (IOEM)
University of Birmingham, Edgbaston, Birmingham B15 2TT, UK
Tel: +44 (0)121 414 6030   Fax: +44 (0)121 414 6217

email: J.B.Grainger@bham.ac.uk Website: www.pcpoh.
bham.ac.uk/ioem

Institute of Occupational Medicine (IOM)
Research Park North, Riccarton, Edinburgh EH14 4AP,
Scotland, UK
Tel: +44 (0)870 850 5131   Fax: +44 (0)870 850 5132
email: info@iomhq.org.uk   Website: www.iom-world.
org

Institution of Fire Engineers (IFE)
London Road, Moreton-in-Marsh, Gloucestershire
GL56 0RH, UK
Tel: +44 (0)1608 812 580   Fax: +44 (0)1608 812 581
email: info@ife.org.uk   Website: www.ife.org.uk

Examination Board

National Examination Board in Occupational Safety and Health (NEBOSH)
Dominus Way, Meridian Business Park, Leicester
LE19 1QW, UK
Tel: +44 (0)116 263 4700   Fax: +44 (0)116 282 4000
email: info@nebosh.org.uk   Website: www.nebosh.
org. uk

Associations and Organisations

Asbestos Removal Contractors Association (ARCA)
ARCA House, 237 Branston Road, Burton upon Trent,
Staffordshire DE14 3BT, UK
Tel: +44(0)1283 531126   Fax: +44 (0)1283 568228
email: Form on website   Website: www.arcaweb.
org.uk

British Red Cross

UK Office, 44 Moorfields, London EC2Y 9AL, UK
Tel: +44 (0)870 170 7000   Fax: +44 (0)20 7562 2000
email: information@redcross.   Website: www.redcross.
org.uk   org.uk

British Safety Council (BSC)

70 Chancellors Road, London W6 9RS, UK
Tel: +44 (0)20 8741 1231   Fax: +44 (0)20 8741 4555
email: Form on website   Website: www.britishsafe-
tycouncil.co.uk

Council of Registered Gas Installers (CORGI)

1 Elmwood, Chineham Park, Crockford Lane, Basingstoke,
Hants RG24 8WG, UK
Tel: +44 (0)870 401 2200   Fax: +44 (0)870 401 2600
email: enquiries@corgi-gas.com   Website: www.corgi-gas.
com

Fire Protection Association (FPA)

London Road, Moreton-in-Marsh, Gloucestershire
GL56 0RH, UK
Tel: +44 (0)1608 812 500   Fax: +44 (0)1608 812 501
email: fpa@thefpa.co.uk   Website: www.thefpa.co.uk

Royal Institute of Public Health (RIPH)

28 Portland Place, London W1B 1DE, UK
Tel: +44 (0)20 7580 2731   Fax: +44 (0)20 7580 6157
email: examinations@riph.org.uk   Website: www.riph.org.uk

Royal Society for the Prevention of Accidents (RoSPA)

RoSPA House, Edgbaston Park, 353 Bristol Road, Edgbaston,
Birmingham B5 7ST, UK
Tel: +44 (0)121 248 2000   Fax: +44 (0)121 248 2001
email: help@rospa.com   Website: www.rospa.org.uk

Royal Society for the Promotion of Health (RSPH)
38A St. George's Drive, London SW1V 4BH
Tel: +44 (0)20 7630 0121   Fax: +44 (0)20 7976 6847
email: rsph@rsph.org   Website: www.rsph.org

Safety and Reliability Society (SARS)
Clayton House, 59 Piccadilly, Manchester M1 2AQ, UK
Tel: +44 (0)161 228 7824   Fax: +44 (0)161 236 6977
email: info@sars.org.uk   Website: www.sars.org.uk

St John Ambulance
UK Office, 27 St John's Lane, London EC1M 4BU, UK
Tel: +44 (0)8700 10 49 50   Fax: +44 (0)8700 10 40 65
email: Form on website   Website: www.sja.org.uk

Appendix E: Industries – principal legal requirements

All industries

Employers' Liability (Compulsory Insurance) Act 1969
Health and Safety at Work etc Act 1974

Building Regulations 2000
Children (Protection at Work) Regulations 1998
Control of Asbestos at Work Regulations 2002
Control of Noise at Work Regulations 2005
Control of Substances Hazardous to Health Regulations 2002
Control of Vibration at Work Regulations 2005
Dangerous Substances and Explosive Atmospheres Regulations 2002
Health and Safety (Consultation with Employees) Regulations 1996
Health and Safety (Display Screen Equipment) Regulations 1992
Health and Safety (First Aid) Regulations 1981
Health and Safety (Miscellaneous Amendments) Regulations 2002
Health and Safety (Safety Signs and Signals) Regulations 1996
Lifting Operations and Lifting Equipment Regulations 1998
Lifts Regulations 1997
Management of Health and Safety at Work Regulations 1999
Manual Handling Operations Regulations 1992
Personal Protective Equipment at Work Regulations
Pressure Equipment Regulations 1999
Pressure Systems Safety Regulations 2000
Provision and Use of Work Equipment Regulations 1998
Regulatory Reform (Fire Safety) Order 2005
Reporting of Injuries, Diseases and Dangerous Occurrences Regulations 1995
Safety Representatives and Safety Committees Regulations 1977

Work at Height Regulations 2005
Working Time Regulations 1997
Workplace (Health, Safety and Welfare) Regulations 1992

Agriculture, forestry and arboriculture

Food and Environmental Protection Act 1985
Control of Pesticides Regulations 1986

Asbestos

Asbestos (Licensing) Regulations 1983 (as amended)
Control of Asbestos at Work Regulations 2002

Biological products

Biological Products Regulations 2001

Chemical

Explosives Acts 1975 & 1923

Carriage of Dangerous Goods (Classification, Packaging and Labelling) and Use of Transportable Pressure Receptacles Regulations 1996

Chemicals (Hazard Information and Packaging for Supply) Regulations 2002

Dangerous Substances and Explosive Atmospheres Regulations 2002

Dangerous Substances (Notification and Marking of Sites) Regulations 1990

Classification and Labelling of Explosives Regulations 1983

Control of Explosives Regulations 1991

Market and Supervision of Transfers of Explosives Regulations
1993
Notification of New Substances Regulations 1993
Packaging of Explosives for Carriage Regulations 1991

Construction

Construction (Design and Management) Regulations 1994
Construction (Head Protection) Regulations 1989
Construction (Health, Safety and Welfare) Regulations 1996
Work in Compressed Air Regulations 1996

Diving

Diving at Work Regulations 1997
Work in Compressed Air Regulations 1996

Docks

Dangerous Substances in Harbour Areas Regulations 1987
Docks Regulations 1988
Loading and Unloading of Fishing Vessels Regulations 1988

Food and catering

Food Safety Act 1990
Food Hygiene (England) Regulations 2005

Foundries

Control of Vibration at Work Regulations 2005

Gas

Gas Safety (Installation and Use) Regulations 1998
Gas Safety (Management) Regulations 1996
Pipelines Safety Regulations 1996

Genetics

Genetically Modified Organisms (Contained Use) Regulations 2000

Leisure

Activity Centres (Young Persons Safety) Act 1995
Adventure Activities (Licensing) Regulations 1996

Major hazard installations

Control of Major Accident Hazards Regulations 1999
Radiation (Emergency Preparedness and Public Information) Regulations 2001

Mines

Mines and Quarries Act 1954
Coal and other Safety-lamp Mines (Explosives) Regulations 1993
Coal Mines (Owners' Operating Rules) Regulations 1993
Escape and Rescue from Mines Regulations 1995
Management and Administration of Safety and Health at Mines Regulations 1993
Mines (Control of Ground Movement) Regulations 1999

Mines (Safety of Exit) Regulations 1988
Mines (Shafts and Winding) Regulations 1993

Nuclear

Nuclear Installations Act 1965
Control of Major Accident Hazards Regulations
Public Information for Radiation Emergencies Regulations 1992

Offshore

Offshore Installations and Pipeline Works (First Aid) Regulations 1989
Offshore Installations and Pipeline Works (Management and Administration) Regulations 1995
Offshore Installations and Wells (Design and Construction) Regulations 1996
Offshore Installations (Prevention of Fire and Explosion and Emergency Response) Regulations 1995
Offshore Installations (Safety Case) Regulations 1992
Offshore Installations (Safety Representatives and Safety Committees) Regulations 1989

Oil

Borehole Sites and Operations Regulations 1995
Control of Major Accident Hazards Regulations 1999
Pipelines Safety Regulations 1996

Plastics

Petroleum-Spirit (Plastic Containers) Regulations 1982

Pottery

Control of Lead at Work Regulations 2002

Quarries

Quarries Regulations 1999

Radiation

Ionising Radiations Regulations 1999
Public Information for Radiation Emergencies Regulations 1992
Radiation (Emergency Preparedness and Public Information) Regulations 2001

Railways

Level Crossings Regulations 1997
Railway Safety (Miscellaneous Provisions) Regulations 1997
Railway Safety Regulations 1999
Railways (Safety Case) Regulations 2000, 2001 & 2003
Railways (Safety Critical Work) Regulations 1994

Service Sector

Care Homes Regulations 2001

Transportation

Carriage of Dangerous Goods by Rail Regulations 1996
Carriage of Dangerous Goods by Road Regulations 1996

Carriage of Dangerous Goods (Classification, Packaging and Labelling) and Use of Transportable Pressure Receptacles Regulations 1996

Transportable Pressure Vessels Regulations 2001

Transport of Dangerous Goods (Safety Advisers) Regulations 1999

Chemicals (Hazard Information and Packaging for Supply) Regulations 2002

Dangerous Substances and Explosive Atmospheres Regulations 2002

Market and Supervision of Transfers of Explosives Regulations 1993

Index